Mentality and Machines

Mentality and Machines

Second Edition

Keith Gunderson

University of Minnesota Press. Minneapolis

First edition published by Doubleday, Anchor Books
Second edition published by the University of Minnesota Press,
2037 University Avenue Southeast, Minneapolis MN 55414
Printed in the United States of America

Library of Congress Cataloging in Publication Data

Gunderson, Keith
Mentality and machines.

Bibliography: p.
Includes index.
1. Intellect. 2. Problem solving. 3. Artificial intelligence. I. Title.
BF431.G844 1985 001.53'5 85-973
ISBN 0-8166-1362-1 (pbk.)

The following chapters are reprinted with permission of the publishers: Chapter One, "Descartes, La Mettrie, Language, and Machines," was originally published in *Philosophy* (Vol. 39, No. 149, 1964, pp. 193-222), copyright © Cambridge University Press; Chapter Two, "The Imitation Game," first appeared in *Mind* (Vol. 73, N.S., No. 290, 1964, pp. 234-45), copyright © Oxford University Press; and Chapter Three, "Robots, Consciousness, and Programmed Behavior," was originally published in *British Journal for the Philosophy of Science* (Vol. 19, 1968, pp. 109-22), copyright © British Society for the Philosophy of Science.

For

PAUL BENACERRAF

and

HILARY PUTNAM

CONTENTS

Preface to the Second Edition

Mentality and Machines, together with the second edition's "Unconcluding Philosophic Postscript," is probably best characterized as a general essay in the philosophy of mind oriented to philosophical and psychological questions about real as well as imagined robots and machines. At times the discussion focuses on actual research projects in the field of artificial intelligence (AI) and what used to be called computer simulation of cognitive processes (CS). (A discussion of some contrasts between past and present nomenclatures is contained in the first major section of the Postscript, and readers new to this book but conversant with recent philosophical debates concerning AI may want to read that first.) But in many places only imagined robots or machines and their possible capacities and behaviors mark the points of departure. And most of the theses that derive from these "thought-experiments" could no doubt have been formulated and argued for independently of any extant machines whatsoever. Overall it seems accurate to say that the main body of the book, except for Chapter Four and parts of Chapter Five is not as closely correlated with the details of research projects in AI as is, for example, Hubert L. Dreyfus' *What Computers Can't Do* (revised edition, 1979). Nevertheless, these projects and critical reactions to them (philosophical and other) have also provided jumping-off points for a number of arguments developed in the Postscript.

Certainly one of the livelier interdisciplinary subjects to which philosophers have addressed themselves in recent years has been that involving the attempt to model or explain the human mind through the use of computer programs (AI). Throughout the 1970s and on into the present researchers in AI have continued to provide copious examples of such projects more or less continuous with their conceptual counterparts of the 1950s and 1960s. And philosophical reactions—both pro and con—to these models with their accompanying controversial glosses have kept pace, adding to the revisions of revisions of attitudes and assessments concerning what all this mix of mentality with machines really adds up to. With the abundant literature of the last decade and a half almost oppressively in mind, some of which surely impinges on portions of *Mentality and Machines*, a gesture of justification seems called for on behalf of the reappearance of a book first published in 1971 (with some of its contents having been published in an article form long before that). Roughly it is this: Chapter One is primarily historical and addressed to a set of seventeenth- and eighteenth-century controversies about human and animal minds and their possibly mechanistic explanations which still seem to me interesting in their own right as well as germane to current disputes. Chapter Two is a critique of A. M. Turing's celebrated answer to the question "Can machines think?" (in his "Computing Machinery and Intelligence"). For a variety of reasons, Turing's influence regarding these issues just does not quit and continues to generate some of the more animated methodological debates within the field of AI (see John Searle's "Minds, Brains, and Programs" and the thousand and one reactions to it in *The Behavioral and Brain Sciences*, Vol. 3, 1980). Chapters Three, Four, and Five explore a number of interrelated conceptual issues—e.g., whether being programmed to do certain tasks is at odds with *really* doing them—and still seem to me relevant to our general understanding of human, robot, and computer capacities and behaviors, as does the most controversial distinction developed in the book—that between program-receptive and program resistant aspects of mentality.

Chapter Four is probably the most antiquated part of the essay since it focuses on early (1950s–1960s) projects in computer simulation of cognitive processes (CS). Even so, many of the very same methodological migraines diagnosed in that chapter as having afflicted researchers attempting to assess in some objective fashion those earlier prototypic machine models of human minds are still troubling AI researchers today.

Revisions for this second edition have not involved any extensive rewriting of the original chapters, but only minor nonsubstantive changes. This is not because I still feel committed to each and every point made in those chapters; much less do I have an attachment to all its stylistic strategies and turns of phrase. But to attempt revisions at that level of discontent now would, I think, result more in a tedious act of author-purification than anything that would interest a reader. And it seemed both easier and more effective to indicate some of my afterthoughts, vacillations, and revised feelings concerning the whole kit and caboodle in a postscript.

The Postscript, it should be emphasized, is not designed to survey or even summarize the Heraclitean flow of machine-intelligence research and the philosophizing about it that has bubbled on since 1971. To have done so would have demanded text at least as lengthy as the book itself, whose major virtue is perhaps brevity. I hope, in fact, to accommodate some of the issues this new tidal wave of material churns up in an altogether new work (in progress) entitled *Interviews with a Robot*. So the updating represented in the Postscript is selective, containing discussions of and digressions on more recent writings both within and outside AI that seemed to have special pertinence to the salient themes of the book. The Postscript also proposes some entirely new distinctions which I hope strengthen and supplement the central arguments of the main text, and includes among them refinements of the previously mentioned dichotomy between program-receptive and program-resistant aspects of the mind.

I would like to think this new edition may have the inadvertent pedagogical virtue of illustrating some of the continuity in philosophical concerns and strategies stretching between the appearance of Alan Ross Anderson's anthology *Minds and Machines* (1964) and John Haugeland's *Mind Design* (1981), which Haugeland has aptly described as a successor to the Anderson volume. Chapter Two of *Mentality and Machines* ("The Imitation Game") appeared in a slightly different form in the Anderson anthology (as well as elsewhere), and a number of remarks and lines of argument in the Postscript happen to use themes in articles contained in Haugeland's volume.

Some of the authors whose writings figure prominently in the Postscript include John Haugeland, Allen Newell, Herbert A. Simon, Zenon Pylyshyn, Hubert L. Dreyfus, Daniel C. Dennett, John R. Searle, and Paul M. Churchland. Except for Churchland, all the above-mentioned contributed chapters to Haugeland's anthology, so I have used the pagination from that volume in my references to them.

The Postscript incorporates (with revisions) a few paragraphs from my "*Content and Consciousness* and the Mind-Body Problem," *Journal of Philosophy*, Vol. LXIX, No. 19, Oct. 5, 1972, pp. 591–604, and "Paranoia Concerning Program-Resistant Aspects of the Mind, and Let's Drop Rocks on Turing's Toes Again," one of many comments on Kenneth Mark Colby's "Modeling a Paranoid Mind" in *The Behavioral and Brain Sciences*, Vol. 4, 1981, pp. 515–60.

I wish to thank my wife, Sandra Riekki, for typing up the really crummy earlier versions of the Postscript, and Sally Lieberman of the Minnesota Center for Philosophy of Science, for her splendid efficiency in wrestling the final felt-tipped, marginalia-marked version of it into and out of the word processor.

I am also indebted to my colleague Gene Mason for a piece of stylistic advice, and to Lindsay Waters, formerly of the University of Minnesota Press, for his sympathetic proddings and encouragement.

Preface to the First Edition

I

The sticky issues involved in characterizing the mental form a philosopher's tar baby to which many fists are fastened. The point of minding about machines lies simply in the hope that Br'er Robot might prove more formidable an opponent than Br'er Rabbit.

Not all philosophers have shared or now share this vision, of course. And not all who have shared the reluctance to share it have shared each other's alternative outlook. Rudeness to robots has been practiced from René Descartes through Gilbert Ryle, who, in *The Concept of Mind,* cautions us that "Man need not be degraded to a machine by being denied to be a ghost in a machine" (p. 328). In the face of eclectic antagonism the perennial popularity of mechanistic analogs of the mind seems all the more fascinating. What explains and how should we assess the philosophical relevance of this *idée fixe,* and what attractions and limitations attend the current art of robotology?

As a path to philosophical enlightenment there may be no *a priori* reason to prefer analyzing analogies between minds and machines to analyzing analogies between minds and mud. *A posteriori,* however, there is much to recommend the former. Mud is as it always was, and as it was and is it holds no particular promise as a medium for modeling the mind. Machines, on the other hand, have, since the seventeenth century, dealt us diachronic delights. Through a process of unnatural selection the Cro-Magnon robot has at last evolved, equipped with enough self-adaptive, problem-solving prowess to compel the collective concern of philosophers, psychologists, and miscellaneous camp followers.

Computer science in the twentieth century has forced philosophers to redraw the line between men and machines in a variety of respects and has thereby given at least a suburban status to what seemed to be the time-hollowed credos of mechanistic materialism. Cybernetics has thus had much the same force that the technical innovations of the Swiss clockmakers had in the seventeenth and eighteenth centuries when their compelling mechanisms inspired a new surveying of the boundaries between the living and the mechanical. From the vantage point of this century, it is all too easy to belittle the philosophical surprise that Jacques de Vaucanson's mechanical duck and flute-playing boy held for his contemporaries who were inclined to associate the movements involved in ambulation only with living organisms. But that these inventions gave impetus to Julien La Mettrie's mechanistic speculations, which in turn became a guidebook for later French materialists such as the philosophes, Pierre Cabanis, *et al.*, is now a matter of historical record.

Indeed, the most important, though casual and almost inadvertent, implications that computer science has had for philosophy are the new insights it has given us into some of our old concepts. Some examples: that a purely mechanistic system can display some degree of self-adaptiveness; that certain highly complicated problem-solving activities can be carried out by indisputably non-living machines; that certain kinds of creative abilities are not at odds with rule-governed behaviors; etc.

II

Nevertheless, the actual import and substantive scope of the imputed insights have remained obscure, and a variety of philosophical perplexities have persisted in pestering our conceptual consciences. Some favored phrasings of these bafflements include: "Can machines think?" "Could a robot have feelings?" "Could it act responsibly?" In 1950 the British mathematician-logician A. M. Turing published his influential "Computing Machinery and Intelligence" in *Mind* and proposed answers to a number of these questions. Since that time, with guppy-like fertility, articles have begotten articles that have begotten articles. At first glance this astonishing proliferation of writings on such matters and mentals appears indicative of an unmitigated metaphysical and psychological bonanza. After further exposure to the awesome output, however, one, through the aid of his mysterious recognition capacities (see Chapter Four), begins to detect rather repetitive patterns of assumption and inference. And at this juncture it is hard not to wonder whether the collectively authored corpus on philosophy and cybernetics is most accurately portrayed as a healthy harvest of new insights or, instead, as a swarm of locusts that threaten the crops. The main type of argument criticized in Chapter Two and the main type of argument criticized in Chapter Three were selected because each represents one of the two dominant positions on minds and machines in the philosophical literature since 1950. These two positions are competitive with each other, but the drift of my own arguments is that both are mistaken and that failure to see why neither approach is viable has hampered an adequate appreciation of the actual analogies and disanalogies between minds and machines.

Simply to avoid a couple of wrong turns on the road to such appreciation, however, hardly brings us to our destination. And perhaps any final arrival is no more to be hoped for than zooming up to that ever-distant mirage of a puddle on the highway ahead. Still, we do get somewhere in driving at it. In the case of pursuing analogies between minds and machines the philosopher should, I think, apprise himself of some of the twists and turns of research in the areas known as "Artificial Intelligence" (AI) and "Computer Simulation of Cognitive Processes" (CS).

But this is no simple task. For the at least numerically impressive philosophical outpouring mentioned above has, amazingly enough, been more than matched in volume by the overflow of books and articles of a more technical or semitechnical nature in the overlapping areas of AI and CS. Results from both these areas, but especially CS, are current causes of theoretical and methodological innovations in various branches of the social sciences, and many aspects of the research have as well philosophical overtones.

The aim of research in AI was not initially directed toward an imitation of the human mind with machines. The goal instead was to imitate by machine what suitably embodied minds can do or bring about. Research in CS, on the other hand, has involved a conscious attempt to construe human cognitive processes in terms of elementary information processes that may be represented by a "trace," or history of moves involved in the execution of certain sorts of computer programs. These techniques have themselves, of course, quickly become items of metamethodological interest and hence tempting subjects for philosophical appraisal. Indications of this are provided by writings such as Kenneth Sayre's *Recognition: A*

Study in the Philosophy of Artificial Intelligence (1965) and (to a lesser extent) his *Consciousness: A Philosophic Study of Minds and Machines* (1969) and Hubert L. Dreyfus' *Alchemy and Artificial Intelligence* (1965) and *What Computers Can't Do* (forthcoming).

Obviously it is not the case that, having failed to extract the desired answers directly from the informal philosophical literature, one can simply turn to the factual oracle of computer science (AI and CS) and find illumination. For the oracle says different things at different times, and sometimes it mumbles. There is considerable disagreement among researchers in these fields themselves, and there is certainly no hearty concensus of metacriticism forthcoming from the writings of philosophers (such as Sayre and Dreyfus) who have attempted to provide us with an acceptable overview.

Part of the purpose of this book, then, after some historical antecedents have been examined, is to extract from the array of contemporary approaches to mentality and machines, some semblance of a uniform perspective from which to assess the bearing that such inquiries have on the philosophy of mind. In no sense is what follows geared to grinding out a simple "yes" or "no" answer to such questions as "Can machines think?" or "Could robots have feelings?" Instead I wish to illustrate why the most favored answers to such queries are mistaken and what this shows us about the mind. I shall also propose that some of the questions that led to these answers might be usefully replaced by questions more closely connected with research in CS and AI as well as to work in biosimulation, an area about which, unfortunately, I say next to nothing. This is, of course, only after certain methodological mishaps in these areas have been circumscribed and cauterized.

III

A word of caution seems in order in regard to the philosophical positions endorsed in the first three chapters. In Chapter One I reconstruct and defend an argument which I believe Descartes used against the view that animals possess considerable mental prowess. In Chapter Two I rephrase this argument and direct it against the position of Turing which would lead us to believe that current computing machines have the capacity for thought. I am interested in Turing's position not only because of what I take to be its mistaken philosophical conclusions, but because these conclusions have been thought acceptable and useful to various researchers in AI and CS. Still, should my objections to Turing be sustained, it would not follow that I have shown that machines do not or will not have cognitive capacities. All that follows is that a certain argument for the conclusion that they do or will possess such capacities is discredited. So, too, if my defense of Descartes is correct, it does not follow that animals are mere machines devoid of thought and feeling. It only shows that a certain argument to prove that they were endowed with a rich variety of mental capacities is discredited. I emphasize these limits on my polemics since they have already been misunderstood. On the basis of my defense of Descartes' argument against the view that animals possess thought and feeling I have been described by one writer as endorsing an antimaterialist position. Keith Campbell, in his article "Materialism" for the *Encyclopedia of Philosophy,* edited by Paul Edwards (New York, 1967, pp. 180–88) claims that I "revived an argument of Descartes' to the effect that men are not machines, even cybernetic machines, and therefore not merely material." But all I attempted to show in "Descartes, La Mettrie, Language, and Machines" was that Descartes formulated an effective argument against a certain type of strategy for proving that animals possessed a rich assortment of mental capacities and that Descartes' argument could also be employed against a certain type of modern strategy for proving that computers are intelligent. Were I correct in my re-employment of Descartes' argument, it would hardly imply that machines (or animals) were not intelligent; nor would it imply that we were not machines of some (perhaps cybernetic) sort. It would *only* follow that a certain strategy for showing that we were was ineffectual.

A similar misunderstanding attends James E. Tomberlin's review of *Minds and Machines,* edited by Alan Ross Anderson (Englewood Cliffs, N.J., 1964). In *Philosophy and Phenomenological Research* (Vol. XXVI, Dec. 1965, No. 2, pp. 278–79), Tomberlin suggests that in "The Imitation Game" I advance three arguments for holding that machines cannot think. But in fact I advance no arguments there for believing that machines cannot think, but only arguments against accepting a certain strategy for proving that they can.

Chapter Three in a superficial sense is a shift of allegiances. For my concern there is with illustrating the inadequacy of a certain type of argument for showing that robots or machines never could have feelings, act responsibly, etc. (Interestingly enough, Descartes endorsed a version of this argument which La Mettrie criticized. So in this chapter my historical preferences are just the reverse of what they were in Chapter One where I defended Descartes and criticized La Mettrie.) I do not, it must be emphasized, purport to prove that robots or machines do or could have feelings, act responsibly, etc. I only attempt to impugn a certain influential type of argument for proving that they couldn't and strongly suggest that *no* philosophical argument could ever show that they couldn't.

The major misunderstanding likely to attend a reading of Chapter Five concerns the distinction I develop there between what I call *program-resistant* features and *program-receptive* features of mentality. Let me simply forewarn the reader that my commitment to this distinction in no way implies a belief that there are *any* features of the mental forever beyond the pale of mechanization. What is implied instead is that there is a host of features that are not susceptible to mechanization through anything like current programming techniques. And I do not, it must be emphasized, identify the possibility of being mechanized with the possibility of being programmed.

The chapters of this book were, for the most part, originally written as chapters for this book. Personal impatience and professional propellants, however, led me to let some of them appear in public in shortened and, perhaps, somewhat less loveable form as articles in philosophical journals and anthologies. Incorporations of my reactions to reactions to their debut in that form has, I hope, helped polish their performance in the finale.

I mention the above for reasons of vanity; it is the custom to ask or be asked, "Is it a book you are publishing, or *merely* a collection of your articles?" (And why "merely"? *My* articles!) Still, I wish to answer too (honestly), "A book. Hah!" Yet accuracy prods me to point out that most of these chapters were originally dechapterized so that they could appear as articles, and did, and have since been de-articleized in order to be presented in the form in which they were originally conceived.

Chapter One, "Descartes, La Mettrie, Language, and Machines," originally appeared under that title as an article in *Philosophy* (Vol. XXXIX, No. 149, pp. 193–222, 1964). It has undergone few revisions, though a section of it has been shifted to Chapter Two. That chapter, "The Imitation Game," was "prematurely" anthologized in *Minds and Machines* (pp. 60–71) and subsequently appeared in *Mind* (Vol. LXXIII, N.S., No. 290, Apr. 1964, pp. 234–45) where it had first been accepted for publication. I hope I have remedied some of the shortcomings attendant upon its initial printing. Chapter Three, "Robots, Consciousness, and Programmed Behavior," enabled me to accept kind invitations for a couple of years from a number of universities to appear as a colloquium speaker. (It may be of some interest to record that during my extensive travels throughout these United States I found no regional concentration of prejudice against robots. It was, instead, uniformly distributed.) After I had actually written something else (e.g., Chapter Four), Chapter Three came to rest, happily, in the *British Journal for the Philosophy of Science* (Vol. 19, 1968, pp. 109–22). Certain sections that were conveniently excised for its appearance there have sneaked back in. "Philosophy and Computer Simulation," Chapter Four, was first printed in *Ryle,* edited by George Pitcher and Oscar P. Wood (New York, 1970). But research leading up to it was begun long ago. (So long ago that although time has not withered my appreciation, the number of the National Science Foundation summer research grant that made part of it possible has been totally forgotten.) Its opening paragraphs have been made the opening paragraphs of this Introduction while a number of substantive sections have been added to it. The closing section of "Philosophy and Computer Simulation" has become a section of Chapter Five, "Some Mental Limitations of Some Machines," the rest of which is brand new. The Epilogue has been largely gleaned from a few pages in "Cybernetics and

Mind-Body Problems," which was published in *Inquiry,* (Vol. 12, 1969, pp. 406–19).

Ancestors of some of these chapters inhabited a Ph.D dissertation written for Princeton University in 1963. Since that time they have been quietly interred in the microfilm morgue at the University of Michigan in Ann Arbor.

I wish to thank the editors of the above mentioned journals and anthologies for permission to reprint here writings that were initially published by them.

Mentality and Machines

CHAPTER ONE

Descartes, La Mettrie, Language, and Machines[1]

I

In *L'Homme machine* Julien La Mettrie at one point discusses the possibility of teaching an ape to speak, and later he suggests that just as the inventor Vaucanson had made a mechanical flute player and a mechanical duck, it might be possible some day for "another Prometheus" to make a mechanical man which could talk. With regard to the former he writes:

> Among animals, some learn to speak and sing; they remember tunes, and strike the notes as exactly as a musician. Others, for instance the ape, show more intelligence, and yet can not learn music. What is the reason for this, except some defect in the organs of speech? But is this defect so essential to the structure that it could never be remedied? In a word, would it be absolutely impossible to teach the ape a language? I do not think so.[2]

(He then proceeds to describe just how he would carry out such a project.)

And with regard to the talking mechanical man he writes:

[1] I wish to thank Leonora Rosenfield for some helpful suggestions in connection with this chapter, though she is of course in no way responsible for what I have written.

[2] Trans. by Gertrude Carman Bussey as *Man a Machine* (published together with the French text of a Leyden edition of 1748), p. 100.

> It (thus) appears that there is but one (type of organisation) in the universe, and that man is the most perfect (example). He is to the ape, and to the most intelligent animals, as the planetary pendulum of Christian Huyghens is to a watch of Julien Leroy. More instruments, more wheels and more springs were necessary to mark the movements of the planets than to mark or strike the hours; and Vaucanson, who needed more skill for making his flute player than for making his duck, would have needed still more to make a talking man, a mechanism no longer to be regarded as impossible, especially in the hands of another Prometheus.[3]

I wish to assess the above remarks in relation to seventeenth- and eighteenth-century discussions of René Descartes' doctrine of the *bête machine*–the view that animals other than man are pure machines devoid of thought or feeling.

Descartes knew, as did disciples such as Malebranche, that one cannot consistently accept certain aspects of Cartesian mechanism (the *bête machine* doctrine) without accepting certain other non-mechanistic aspects of Cartesian metaphysics. I shall attempt to show that this was at best imperfectly understood by La Mettrie, and has at times not been adequately accounted for by commentators on the *bête machine* controversy. In order to do this I shall first discuss two ways in which Descartes thought men could be distinguished from any conceivable machines, and why he thought this showed that men differed radically from beasts with respect to rational activities. I shall argue that by developing these claims together with supplementary arguments and assumptions, Descartes did propose a valid argument against a certain view (typical of Montaigne and others) that animals are able to think and reason. I wish to comment on this argument in some detail and rephrase it in a contemporary

[3] *Ibid.*, pp. 140–41.

idiom, *for I am not only interested in its role in the bête machine doctrine, but should like also to illustrate the force I believe it has if used against a certain type of argument to the effect that modern-day computing machines are able to think and perform intelligently.* (I do this in Chapter Two.) La Mettrie, however, uses different arguments to support his mechanistic conclusions, and if certain of his assumptions are correct, these arguments would enable him to circumvent the kind of Cartesian objection which is effective against the Montaigne-type reasoning. But I shall try to show that La Mettrie's way of arguing leads to serious contradictions when taken together with other statements in *L'Homme machine.* And in passing, I hope to show that La Mettrie's interest in talking apes and a talking mechanical man, frivolous and rhetorical as it may seem at first, was no doubt a serious one derived from his concern with problems in the philosophy of mind, and that there are good reasons to suppose that what he had in mind when he imagined the mechanical man was nothing less than a rational being whose intelligence differed not in kind, though perhaps in degree, from the intelligence of a man. This is in direct disagreement with Friedrich Lange who in *The History of Materialism* has written:

> It certainly must not be supported that by a speaker La Mettrie had meant here a rational man; yet we see how delighted he is to compare the masterpieces of Vaucanson, which are so characteristic of their age, with his human machine.[4]

It should also be mentioned at the outset, that in spite of the logical problems involved in any extension of the *bête machine* to man, mechanism did play a very promi-

[4] Trans. by Ernest Chester Thomas, Book I, Section 4, Chap. 1, p. 75. Also, see Aram Vartarian, *La Mettrie's L'Homme Machine,* pp. 19–20.

nent role in Descartes' philosophy of mind. With the exception of introspection and the *cogito* argument, the workings and "movements" of the soul or substantial self were always explained and referred to by physiological and speculative neurological terms (the pineal gland and the animal spirits). In other words, the so-called connection between the soul and what Descartes regarded as the body-machine was almost always accounted for in a mechanistic manner. This pervasive "physiologizing" of the soul together with the *bête machine* doctrine composed the mechanistic half of Cartesian dualism, and in many ways it was this mechanism which became the immediately influential part of Descartes' metaphysics.

In contemporary philosophy, however, especially since Professor Ryle's *Concept of Mind,* it has become increasingly popular to display Descartes' doctrine of the mind as the paradigm of a "Ghost in the Machine" dualism (emphasis *always* on the "Ghost"), and to do so in such a way that virtually anyone capable of being interested in it in the first place will find himself talented enough to devise a refutation of it. It has been almost completely forgotten by contemporary philosophers (though not by historians of ideas) that mechanistic aspects of Cartesianism came to wield a major influence during the century following Descartes' death. At the end of the nineteenth century, however, T. H. Huxley paid elaborate tribute to Descartes' mechanism as a forerunner of his own physiological and biological theories.[5] And it is somewhat ironic that the main outline and outlook of much of Cartesian mechanism is almost identical with the main outline and outlook of philosophical positions

[5] See his "On the Hypothesis that Animals are Automata, and its History" (1874) and "On Descartes' 'Discourse Touching the Method of Using One's Reason Rightly and of Seeking Scientific Truth'" (1870), reprinted as Chapters V and IV in *Method and Results,* Vol. I in the *Essays.*

which in this century are regarded as essentially anti-Cartesian. It is no exaggeration to say that many mechanistic features of Descartes' philosophy of mind have much in common with various physicalistic and behavioristic views which have been supported by the logical positivists, with certain philosophical claims which have issued from cybernetics, and even with the type of mind/body identity thesis currently supported by Professors Feigl, Smart, and others.[6] To say this of course is not to deny that there was a Ghost in Descartes' machine. It is simply to call attention once again to the fact that there was the machine, and that many believed that the Ghost disappeared in the gears. It is to give Descartes full credit for both halves of his dualism.

So we shall now turn to specific features of the *bête machine* controversy—the discussion of the animal soul—which may be viewed as a seventeenth- and eighteenth-century counterpart of virtually the whole cluster of recent discussions which have taken place on the topic of mentality and machines.[7]

[6] As contained in Herbert Feigl's "The 'Mental' and the 'Physical,' " in *Minnesota Studies in the Philosophy of Science,* Vol. II, Eds. Herbert Feigl, Michael Scriven, and Grover Maxwell, pp. 370–497. See especially Feigl's interesting and sympathetic rendering of Descartes' *bête machine* doctrine, p. 412. However, it is Descartes' physiologizing of the soul and not the *bête machine* doctrine which give rise to the similarities with Feigl's monism. Also see J. J. C. Smart's "Sensations and Brain Processes," *Philosophical Review,* LXVIII (1959), pp. 141–56, and his paper "Materialism," *The Journal of Philosophy,* Vol. LX, No. 22: Oct. 24, 1963, pp. 651–62.

[7] That is to say those questions which have been discussed since 1950 in various symposia, books, and to a greater extent in philosophical journals, such as "Can machines think?," "Are computers intelligent?," "Could robots feel?," "Are we after all nothing but a sort of robot or machine? or would certain sorts of robots or machine after all be nothing but a sort of us?," *et al.* To pose

and answer some of these questions in mechanistic philosophy *might,* of course, help to answer questions concerning topics such as other minds, behaviorism, free will, and the mind/body relationship—and vice versa.

Similarly, in the seventeenth and eighteenth centuries virtually identical questions arose because of Descartes' claim to have proved that animals were pure machines. One could quarrel with his claim in a variety of ways: for example, by denying that animals were machines, quibbling with Descartes' use of the word "machine," or by denying that because they were, they were thereby deprived of thought and feeling. This naturally gave rise to the question as to whether or not machines could be said to think, and whether since we undoubtedly think, we could be said to be mere machines, and so forth. So good is the parallel between the *bête machine* controversies and recent discussions of mentality and machines, that I believe almost every major position which has been taken with respect to the current discussions has a counterpart in the seventeenth and eighteenth centuries, with the exception of those connected with certain studies in computer simulation of cognitive processes and artificial intelligence and anti-mechanistic arguments based on metamathematical theorems such as Gödel's Incompleteness Proof. (In connection with the latter topic see, for example, J. R. Lucas' "Minds, Machines and Gödel," in *Philosophy,* Vol. XXXVI, No. 137, April and July 1961, pp. 112–27; and for a view which conflicts with Lucas' see J. J. C. Smart's "Gödel's Theorem, Church's Theorem, and Mechanism," in *Synthèse,* Vol. XIII, No. 2, June 1961, pp. 105–10.) Even Lucas' complicated argument based on Gödel's theorem which he believes distinguishes human minds from machines may, however, be seen as essentially an argument that machines could never be self-conscious in the way that human minds are, and that Gödel's theorem illustrates this. This has certain affinities with Descartes' proposal that linguistic activities distinguish men from beasts and machines. There are also the obvious technological parallels, namely, that Vaucanson's inventions may be viewed as having had some of the same sort of philosophical suggestiveness that the digital computer has had recently. At any rate I would disagree with Albert G. A. Balz's remarks where he writes concerning the animal soul controversy: "We may look upon the whole controversy, upon its issues, its arguments, and its oppositions of ideas, with complete detachment. Our interest in its analysis is very much like the interest with which we study a game that has been played,

II

Descartes found no difficulty in the admission that man's body could be regarded as nothing more than a machine.[8] God, of course, was regarded as the maker of it, and it is

---

that is over and done with. This incident in the history of ideas we may regard as closed" (in his "Cartesian Doctrine and the Animal Soul: an Incident in the Formation of the Modern Philosophical Tradition," in his *Cartesian Studies,* p. 148). I believe that Balz has failed here to appreciate that questions concerning the animal soul were intimately bound up with more general issues in mechanistic philosophy–issues still very much alive. Cf. Vartanian's *La Mettrie's L'Homme Machine,* pp. 131–36.

George Boas in his book *The Happy Beast in French Thought of the Seventeenth Century* gives an interesting account of Descartes' *bête machine* doctrine viewed as a response to the tradition of Theriophily–the doctrine that animals (which are more natural than man) are in many ways our superior–even in matters involving thought and intelligence; at least they are believed to be our equals in most things. One way to rephrase some of my above claims would be to say that computerophily has simply replaced theriophily, and thereby given rise to discussions on mentality and machines. I shall later argue that one of Descartes' arguments against theriophily was valid, and is also valid today against a certain brand of computerophily. This is, I suppose, one way of partially supporting what might be called a *machine machine* doctrine–or the doctrine that machines are nothing but pure machines.

[8] This points up a difference between the tradition of mechanistic philosophy which involves the topic of mentality and machines and derives ultimately from Descartes' physiologizing and *bête machine* doctrine, and that which is usually referred to as the vitalist/mechanist controversy or the teleology vs. mechanism discussions. Here Descartes may be seen as having already assumed answers to the major issue involved in the latter controversies. He is not generally concerned with distinctions between the animate and inanimate.

> incomparably better arranged, and possesses in itself movements which are more admirable, than any of those which can be invented by man.

He suggests, however, that if we were able to make machines

> possessing the organs and outward form of a monkey or some other animal without reason, we should not have had any means of ascertaining that there were not of the same nature as those animals.[9]

But, he claims, that even if we were able to make such machines as would imitate the outward form and organs of our body, there would always be two tests at our disposal by which we could tell whether they were real men or simply machines which looked like men.

The first test is

> that they could never use speech or other signs as we do when placing our thoughts on record for the benefit of others.[10]

For

> it never happens that it arranges its speech in various ways, in order to reply appropriately to everything that may be said in its presence, as even the lowest type of man can do.[11]

The second test is that

> although machines can perform certain things as well as or perhaps better than any of us can do, they infallibly fall short in others, by which means we may discover that they did not act from knowledge, but only from the disposition of their organs. For while reason is a universal

[9]In *The Philosophical Works of Descartes,* trans. by E. Haldane and G. R. T. Ross, from *Discourse on Method*, Vol. 1, p. 116. Unless otherwise indicated, translations from French texts are by Donna Gunderson-Rogers.

[10]*Loc. cit.* [11]*Loc. cit.*

> instrument which can serve for all contingencies, these organs have need of some special adaptation for every particular action. From this it follows that it is morally impossible that there should be sufficient diversity in any machine to allow it to act in all events of life in the same way as our reason causes us to act.[12]

These observations, then, supposedly enable us (a) to distinguish men from machines, and (b) to recognize one difference which exists between man and beast with respect to linguistic and non-linguistic rational performances: We may call these two tests respectively (1) the language test, and (2) the action test. We may also talk about (1) and (2) as tests which a beast or a machine may pass or fail to pass. (Roughly: if beast or machine is able to perform whatever linguistic or non-linguistic and seemingly rational acts which human beings are capable of, and in such a way that in all respects the actions themselves of the beast or machine are indistinguishable from a human being, we will say that the beast or machine is able to pass the language or action test. If not, we shall assume the beast or machine fails to pass them.) Though Descartes did not use the word "test" in this way, it will simplify matters to talk as though he did. This will in no way change the substance of his remarks.

Descartes does admit that a machine could be constructed so that "if it is touched in a particular part it may ask what we wish to say to it; if in another part it may exclaim that it is being hurt, and so on".[13] But he believes, as the aforementioned quotes have shown, not *all* the replies of such a machine would be appropriate; in fact, though we could rig a machine to make certain "appropriate" responses, there would still remain a wide range of linguistic activities in which its shortcomings would be revealed. In other words, the machine would always fail

[12] *Loc. cit.* [13] *Loc. cit.*

at some point to pass the language test. Now there are no reasons which Descartes offers which would show that these shortcomings could not, within his framework, be construed simply as instances of human actions which machines, as he asserts, would "infallibly fall short" in duplicating. Hence, the language test, as I have called it, is exactly the same *type* of test as the action test, with the exception that it subsumes a more specific (though still very broad) range of activities. Furthermore, in the *Discourse* when Descartes goes on to provide illustrations of animals failing to pass these tests, they are all essentially examples which involve the failure to use language in certain ways and situations, so that we supposedly come to know that the animals really do not reason or have thoughts to communicate. Though he does mention actions in general again, he does not provide examples of ways in which animals fail to emulate us, but simply writes:

> It is also a very remarkable fact that although there are many animals which exhibit more dexterity than we do in some of their actions, we at the same time observe that they do not manifest any dexterity at all in many others.[14]

Consequently, though Descartes distinguishes two tests at the outset, he actually relies more on the one test—the language test—which, if not passed by a certain subject S (who is representative of a class of subjects) would entail S's failure to pass the action test, since given the vagueness and generality of his overall account, the language test may be construed as one way of showing whether or not a subject is capable of a certain (broad) range of actions. For all Descartes needs in order to show that S has not passed the so-called action test is that there is some (broad) range of actions where S (machine or beast, for example) fails to perform in ways comparable to the ways

[14] *Op. cit.,* p. 117.

in which human beings perform. (It would not of course work the other way: failure to pass the action test would not entail failure to pass the language test.)

Descartes himself later came to write that he believed the use of language was the only infallible way of distinguishing men from beasts with respect to rationality. At any rate, twelve years after the *Discourse* was published he wrote to Henry More:

> For language is the one certain indication of latent cogitation in a body, and all men use it, even the most stupid and mentally deranged, and those deprived of their tongue and vocal organs, whereas on the other hand not a single brute speaks, and consequently this we may take for the true difference between man and beast.[15]

And in reply to some of More's objections, he wrote:

> When, however, "dogs nod yes with their tails," etc., these are merely motions which accompany the passions, and as such, I believe, are to be accurately distinguished from language, which alone is certain proof of latent cogitation in the body.[16]

Nonetheless, it is of little consequence here whether or not the majority of Descartes' followers and foes came to know that he regarded the language test more highly than the action test. The emphasis even in Descartes' public writings seems to be on the language test, though I know of no place where he explicitly abandons the action test, and, as we shall see, as late as 1646 (in a letter) he reiterates the same view of non-linguistic animal actions

[15] In Leonora D. Cohen's (later Leonora Rosenfield) "Descartes and Henry More on the Beast-Machine—A Translation of their Correspondence Pertaining to Animal Automatism," *Annals of Science, A Quarterly Review of the History of Science since the Renaissance,* Vol. I, No. 1, January 15, 1936, p. 53. From a letter dated February 5, 1649.

[16] *Ibid.,* p. 55. From a letter dated April 15, 1649.

to which he had subscribed in the *Discourse.* It seems that he simply regarded the former as more certain and important than the other, not the other as having no worth at all. (Together, of course, the tests cover human and animal (and machine) performances *of any rational or seemingly rational sort whatsoever.*) And it may well have been the case that Descartes came to believe that for any example of a non-linguistic but seemingly rational activity which he could cite to illustrate man's superiority over beasts it was always possible to refer to some example of a non-linguistic but seemingly rational activity in which a certain sort of beast would surpass human performances; whereas (ignoring current speculations about porpoises) it would be difficult to cite any cases of purely linguistic performances where beasts even seemingly equalled or surpassed human beings. Furthermore, the obvious physical differences between man and most animals, and among various kinds of animals themselves, would seem to provide a simple way of accounting for the differences in non-linguistic performances which exist between man and most animals, and among the animals themselves, for example, performances which involve running, flying, building and so forth, and depend upon certain muscles, limbs, and the like. Such a simple and obviously physical explanation of why beasts do not do certain things which we do circumvents the need for any alternative explanation, which, as we shall see presently, Descartes does have reason to resort to in the case of linguistic performances–given his assumptions. But for the moment we may simply point out that Descartes believed that if and only if subject S cannot pass test *L* (and *A*) S does not possess "characteristic" r, capacity to reason, think, etc.; beasts (and machines) are subjects which cannot pass tests *L* (and *A*), therefore beasts (and machines) do not possess characteristic r; and, there are certain conceivable machines which can do anything a beast can do, and which would be indistinguishable from them. Now we must turn to the reasons why Descartes thought the initial premise of the above summary was justified. In doing so we shall concentrate on the language test.

In the *Discourse* Descartes argues as follows.[17] It is not due to any lack of physical apparatus that animals will fail on the language test. A parrot or magpie, for example, may be assumed to have the organs necessary to speech, since they are able to produce a great many of our words. He does not argue this point in detail, and, of course, one might suggest that parrots really fail to match human performance even on a purely phonemic level, and that this is due to a lack in physical apparatus. But Descartes does not discuss this and simply generalizes from a rather limited set of speech imitations to the conclusion that beasts possess physiological mechanisms as adequate to the production of speech as those possessed by humans. (It is this kind of physiological parallel which it would be so obviously difficult to establish between man and beast with respect to differences in non-linguistic actions. Hence even given Descartes' other assumptions, these other disparities would not seem to cry out for an explanation of a non-mechanistic type. Consider: "Why don't snakes type letters?" "Because they don't have fingers.") Furthermore, Descartes makes no attempt to assess the magnitude of failure which beasts would exhibit in the use of language or the duplication of rational human actions. But he does make it clear, at least in the case of using language, that the successes of beasts are very limited. It is obvious that he would regard the parrot as having failed to pass the language test not because it failed to initiate a fine oral tradition of parrot poetry, but because it could not even match an idiot in its arrangement of words and the making of statements. Descartes seems to be saying

[17] *Op. cit.*, pp. 116–17.

that verbal success is not guaranteed by physiological apparatus since beasts fail to attain it even when (as in the case of parrots and magpies) they may be assumed to have physiological apparatus comparable to that which in our possession leads to just such success. As a consequence, of course, he argues that our success will need to be accounted for in a non-mechanistic, hence for him dualistic, fashion. The possibility that syntactic-semantic-pragmatic skills, the arranging of "different words together, forming of them a statement,"[18] could be due in part to certain innate though not yet discovered neurophysiological mechanisms, did not occur to Descartes. And if this could in some way be shown along with demonstrated differences between animals and humans with respect to the mechanisms involved, Descartes' doctrine would fail to support the further claim that in the case of human beings the facility with language cannot be explained in a purely mechanistic or materialistic manner.

The way in which Descartes explained the differences in performance between man and beast is well known; we do have knowledge of a thinking substantial self, which, in cases other than sleepwalking, reflex movements, *et al.*, initiates and directs various bodily movements, and expresses verbally or in writing or gesture the thoughts which it possesses. Reason, wonder, reflection, feeling,[19]

[18] *Loc. cit.*

[19] Descartes does admit, however, that beasts are capable of a kind of purely mechanical or "unconscious" feeling and sensation. This notion is left unclarified, though he seems to be suggesting that certain reflex movements would perhaps involve feeling or sensations of a sort; a dog would, supposedly, in some mechanical way react to and feel a swift kick but would not, on Descartes' account, have pain. On the whole I believe Descartes does not argue as effectively for the claim that beasts lack all feeling and consciousness, as he does for the claim that they do not think or reason. Since he generally centers his remarks around the latter claim, we shall too. See his letter to More, February 5, 1649, *op. cit.;* also

and so forth, are all activities which may be attributed to this self working within and usually through the body. The way to account for the lack in the linguistic performances of animals is through a corresponding lack of a substantial self or what in effect turns out to be an immortal soul. The paucity of expressions is to be explained by the total absence of thoughts to express. If there were thoughts, they would be expressed in ways comparable to our own, for there is no physiological barrier to their expression. No thoughts because there is no soul—the essence of which is to think. And lacking this, animals emerge as nothing more than machines. (He does not discuss the possibility of what within a Cartesian framework we might call "a self-suppressed soul"—namely one which just didn't feel like expressing anything.)

This was a tidy and attractive solution to various theological dilemmas of the seventeenth century.[20] If the arguments which were used to show that men had souls also worked to show that gnats, lizards, and vermin had souls, many would begin to suspect the arguments.[21]

Cohen's footnote 9, p. 56, *op. cit.* Cf. Vartanian, *Diderot and Descartes,* p. 210. (Descartes apparently had a dog which he treated as if it were to be treated kindly. Malebranche and others were not above consistently [pun] kicking them.)

[20] Cf. Balz, *op. cit.,* p. 111.

[21] In a very similar vein Hilary Putnam in his article "Minds and Machines" indicates suspicion for certain dualistic arguments where he writes, "To put it differently, if the mind-body problem is identified with any problem of more than purely conceptual interest (e.g., with the question of whether or not human beings have "souls"), then *either* it must be that (a) no argument *ever* used by a philosopher sheds the *slightest* light on it (and this independently of the way the argument tends), or (b) that some philosophic argument for mechanism is correct, or (c) that some dualistic argument does show that *both* human beings *and* Turing machines have souls! I leave it to the reader to decide which of the three alternatives is at all plausible." In *Dimensions of Mind,* ed. Sidney Hook, p. 176.

Descartes' *bête machine* doctrine preserved man's distinctness and dignity, and, of course, fit snugly into his general account of the mind as something distinct from the body (ignoring for the moment his weakness for "physiologising the soul"). For given his doctrine of the thinking self, animals could not be anything other than their bodies–mere machines–unless they too possessed souls or egos. Nothing which they *do* or *say* would lead us to believe that they possess such souls, hence, in the alleged absence of any independent reasons for attributing souls to them, why not conclude that they have none? Furthermore, in a letter to the Earl of Newcastle, Descartes writes:

> One can only say that although beasts do nothing which assures us that they think, at the same time, because the organs of their bodies are not greatly different from ours, one can conjecture that there is some thought connected with these organs, such as we experience in ourselves, although theirs is much less perfect. To which I have nothing to reply, except that, if they thought as we, they would have an immortal soul as well as we; which is not likely, because there is no reason at all to believe it of some animals without believing it of them all, and several of them, such as oysters, sponges, etc. are too imperfect for us to be able to believe that of them.[22]

In such cases Descartes may be seen as having used his belief in the absence of an animal soul to buttress his doctrine of the *bête machine*. It did not matter where Descartes started: (1) with a belief in the absence of the

---

There is little doubt that Putnam would regard arguments such as those in (c) as self-impugning–and for virtually the same sorts of reasons that people would regard arguments which proved that gnats were immortal as being for that reason much less convincing.

[22] *Oeuvres de Descartes,* eds. C. Adam and P. Tannery, IV, p. 576.

animal soul, or (2) with a belief in the *bête machine* doctrine. Each could support the other. Of course, if (1) were the only reason for (2), and *vice versa,* there would be a vicious circularity. But Descartes generally avoided this and brought independent premises in to support one or the other; as for example, in the letter quoted above he suggests that the only cases where beasts surpass us in their behavior are cases analogous to those actions of ours which are not directed by our thoughts, but are purely automatic (such as sleepwalking).[23]

III

But in order to provide clear sailing for the *bête machine* doctrine, Descartes first had to account for the fact that a wide variety of animal achievements had since Montaigne, at least, been used as proof that animals were equal to man in certain respects, and able to surpass him in others.

Whether or not he had Montaigne's views specifically in mind when he formulated the *bête machine* doctrine is of little importance here. In the above letter he did write:

> Because although Montaigne and Charron have said that there is greater difference between man and man than between man and beast, all the same a beast has never been found so perfect that it used some sign to make other animals understand something which had no connection with its passions; and there is no man so imperfect that he does not make use of them; so that those who are deaf and dumb invent special signs, by which they express their thoughts, which seems to me a very strong argument for proving that the reason why beasts do not speak as we do is that they don't have any thoughts and not that they

[23] *Ibid.,* p. 575.

lack voices. And one cannot say that they talk to each other but that we do not understand them; because, as dogs and some other animals express their passions to us, they would also express to us their thoughts if they had them.

I certainly know that beasts do many things better than we do, but I am not surprised because even that serves to show that they act naturally and by springs, as does a watch which tells what the hour is better than our judgment is able to do. There is no doubt that when the swallows come in the spring they are thereby acting as watches do. Everything which the honeybees do is of the same order, and the order which keeps cranes in flight, and that which monkeys observe in fighting, if it is true that they observe any, and finally the instinct to bury their dead, is no more strange than that of dogs and cats who scratch the ground in order to bury their excrement, although they practically never succeed in doing it, which shows that they do it only by instinct and without thinking about it.[24]

So not only do we know that he (1) was aware of Montaigne's views on animal behavior but also that (2) he brought arguments to bear on them which in many respects duplicated those presented in his formulation of

[24] *Ibid.*, pp. 575–76. But in her book *From Beast-Machine to Man-Machine,* Leonora Rosenfield writes: "He knew the position of Montaigne and Charron as we have seen. But unlike his contemporary Chanet, he did not appear primarily motivated by the polemicist's desire to refute these two defenders of animal intelligence. This is evidenced by the fact that he mentioned them only casually, and then not until 1646," *op. cit.,* p. 19. The letter she refers to is undoubtedly the one to the Earl of Newcastle from which we have quoted above. I do not wish to quarrel with Rosenfield's thesis. I assume that her main point is simply that Descartes had more general aims in mind than the refutation of one or more particular writers. But I would emphasize point (2) above, and suggest that since Montaigne's views were so widely known, Descartes may simply at times not have felt it important to mention him, just as many people today discuss the "use theory of meaning" without mentioning Wittgenstein.

the *bête machine* doctrine in the *Discourse.* Furthermore, even if he had not found it necessary to refute Montaignism, (3) during the course of the *bête machine* controversies, defenders such as Malebranche did find it desirable[25] and came to regard the *bête machine* as the last word on the subject of animal intelligence; and (4) if he had not had Montaigne or Montaigne's followers' views in mind when he formulated his doctrine, he did have in mind views which were *just like them,* and was right in seeing that he had to meet the kind of challenge which they presented.

Animal feats such as the swallows returning in the spring and the labors of bees are also mentioned in Montaigne's "Apology for Raimond Sebond," but since we find virtually every imaginable animal performance described in that work, the mere fact that Descartes invoked the same examples in his letter cannot by itself establish that it was Montaigne's "Apology" which he was specifically attacking. Nevertheless, most of Montaigne's examples of the prowess of beasts are on the order of swallows returning in the spring to the same place, spiders weaving their webs, fish trapping their prey in unusual ways, and so forth. For example:

> In the life of the tunnies we may observe a singular knowledge of the three parts of mathematics. With regard to astrology they teach it to man, for they stop at the place where the winter-solstice overtakes them, and do not move from thence until the following equinox; wherefore even Aristotle readily grants them that science. . . .[26]

[25] *Treatise Concerning the Search After Truth,* trans. by T. Taylor, Oxford, 1694, Chap. V, "Of Montaigne's *Book,*" pp. 96–100; and see esp. Bk. II, p. 99. For statements and defences of Descartes' *bête machine* doctrine, see esp. Bk. IV, p. 166, and Bk. VI, pp. 81–82.

[26] *Essays of Montaigne,* trans. by E. J. Trechmann, Oxford, Bk. II, Chap. 12, "Apology for Raimond Sebond," p. 473.

This is followed by similar arguments on behalf of the Euclidean capabilities of the tunnies in geometry and arithmetic. To be sure, there is much irony and wit in Montaigne's lines and this sometimes colors his way of arguing. But the over-all effect which he strives for is to be taken seriously. One of the main points which he wishes to establish by the above sorts of examples is summed up in the following:

> I say then, to return to my theme, that there is no reason to imagine that the beasts do, through a natural and enforced instinct, the same things that we do by choice and skill. From like results we must infer like faculties (and from more abundant results, more abundant faculties). . . .[27]

It is this kind of claim with which Descartes had to contend in order to provide clear sailing for the main conclusions of the *bête machine* doctrine.

He meets the argument head on. He is by no means willing to accept Montaigne's reasoning that from "like results we must infer like faculties" or that "from more abundant results" we should infer "more abundant faculties." On the contrary, Descartes believes that from the fact that certain "like results" are attained by animals which imitate man, we should infer "not merely that the brutes have less reason than men, but that they have none at all,"[28] since (in the cases he alludes to) it is clear that very little is required in order "to utter words just like ourselves."[29] Similarly, he is not at all willing to admit that like or superior results in the cases of animal actions are tenable grounds for claiming that they

[27] *Ibid.,* p. 451.

[28] He is thus not content simply to make the now old-hat point in biology and psychology that likeness in function does not guarantee likeness in structure or process.

[29] The quoted phrases not taken from Montaigne are from the *Discourse,* p. 117.

are as rational or more rational than we are. Rather, he writes:

> Hence the fact that they do better than we do, does not prove that they are endowed with mind, for in this case they would have more reason than any of us, and would surpass us in all other things. It rather shows that they have no reason at all, and that it is nature which acts in them according to the disposition of their organs, just as a clock, which is only composed of wheels and weights is able to tell the hours and measure the time more correctly than we can do with all our wisdom.[30]

I think that Descartes' central reasoning here is fundamentally sound and important.

If Montaigne-type arguments were the only ones available for showing that beasts were intelligent, I believe Descartes' *bête machine* doctrine would be true. Other arguments are, of course, available, and some of these will be discussed when we come to La Mettrie. Nonetheless I do think Descartes' arguments are effective against Montaigne, and may also be used as I shall use them in Chapter Two against a certain sort of argument which has been invoked in an attempt to show that computers can think and perform intelligently.

The point in a thimble is this: if the machine or animal is only capable of doing that one thing better than us, or in a way comparable to the way in which we do it, this is an excellent reason for revising our explanation of how that animal or machine does what it does.

Montaigne's reasoning is even slipperier than Descartes seems to realize. In effect Montaigne argues from the particular skills of particular animals to the general conclusion that animals are able to think, reason, and so forth, which could then be used, for example, as the basis for saying a bird can think. In other words, any one ani-

[30] *Loc. cit.*

mal in effect gets credit for the skills of all other animals. The fox gets credit for the spider's weaving skills, and so forth.

Descartes' further explanation of how the machine (or beast) does what it does is "that it is nature which acts in them according to the disposition of their organs, just as a clock, which is only composed of wheels and weights is able to tell the hours and measure the time more correctly than we can do with all our wisdom."[31]

When we are unable to support a large number of the usual assumptions we would make about a subject who was able to calculate, when we find out about the rigged brain and so forth, we should not then stick to our guns and say, "But the end result is the same, so he must be very good at mathematics, etc." Instead we should simply acknowledge that similar end results can be brought off in very different ways. Why should we simply button up our understanding of a vast variety of situations and treat an isolated case which has a very different possible explanation as on a par with the usual cases? Though admittedly Descartes' brand of argument does not prove that in principle a machine could never be conscious, think, or feel, or that psychology, neurophysiology, or genetics will never reveal that we too are "programmed" in hitherto unexpected ways; it does at least bring out the inadequacy of the Montaigne-type of reasoning and a certain sort of computerophily, when these are considered adequate in themselves for showing that animals or computers are capable of thought and intelligence.

IV

La Mettrie, however, presented a different sort of challenge. There are two ways in which it may be outlined.

[31] *Loc. cit.*

The first is based on *L'Homme machine* as well as on *Les Animaux plus que machines* which was published only three years later, in 1750. The second is based on both the just-mentioned works, as well as on his *Histoire naturelle de l'âme* which was published two years before *L'Homme machine.*[32] It should be noted that the line of reasoning in (I) involves a number of conflicting claims. These I explain later.

> (I) Descartes has proved that animals are pure machines. He is right in this, but it is possible to show that we are not essentially different from the animals, hence we can be shown to be machines too. One way in which we can illustrate this lack of disparity is to show that certain skills and abilities which some have held to be peculiar to man, could also be shared by animals (for example, an ape which was trained to use a language) and by machines (such as a mechanical talking man). This would not perhaps show that they possessed the same degree of intelligence as man, but it would show that there is no *essential* difference between their wits and ours. Our mechanisms simply possess a greater and more subtle degree of organization and complexity. We are after all machines, but machines which can think. By getting animals which Descartes has proved to be mere machines to exhibit linguistic skills, which supposedly separated men from beasts, we find that they are not unlike us in any essential way, hence we too are machines. Hence the *bête machine* transforms itself into the *homme machine.* Descartes was on the right track; we have simply pressed his insight about animals to its logical conclusion.

(II) Descartes has shown that animals are machines, but

[32] I shall continue to use the Bussey translation of *L'Homme machine,* but quotations from *Histoire naturelle de l'âme* and *Les Animaux plus que machines* will be taken from La Mettrie's *Oeuvres philosophiques,* Amsterdam, 1774, 3 vols. ("Nouvelle edition, corrigée & augmentée"). Here the *Histoire naturelle de l'âme* is printed in Vol. I under the title *Traité de l'âme.*

not that they are pure machines devoid of thought and feeling. They are like us–sentient and intelligent creatures–and we are like them, complicated machines which give rise to thought and feeling.

In his *Histoire naturelle de l'âme* one of the points which La Mettrie wishes to emphasize is that Descartes was quite mistaken in thinking that animals were *pure* machines. At the beginning of Chapter VI he refers to this latter claim as a laughable opinion,[33] but at the end of Chapter VII takes greater pains to point out the exact differences between his own views and those of the Cartesians, lest certain important similarities lead people to think there is no difference. He writes:

> But first let me be permitted to reply to an objection which a clever man has made to me. 'You admit, he says, in animals, as a basis for consciousness, no substance which is different from matter: why then treat cartesianism as absurd because it supposes that animals are pure machines? And what great difference is there between these two opinions?' I reply with a single word: Descartes denies all consciousness, all faculty of feeling to his machines, or to the matter of which he supposes that animals are uniquely made: and I prove clearly, if I am not greatly mistaken, that if there is a being which is, so to speak, moulded of feeling, it is the animal; it seems to have received everything in this coin, which, in another sense, so many men lack. There is the difference between the celebrated modern of whom I have just spoken, and the author of this work.[34]

But two years later, in *L'Homme machine,* he writes with respect to Descartes:

> This celebrated philosopher, it is true, was much deceived, and no one denies that. But at any rate he understood ani-

[33] *Oeuvres philosophiques,* Vol. I, pp. 83–84.
[34] *Ibid.,* pp. 89–90.

> mal nature, he was the first to prove completely that animals are pure machines. And after a discovery of this importance demanding so much sagacity, how can we without ingratitude fail to pardon all his errors?[35]

It might at first seem, then, that La Mettrie had completely changed his mind about animals since writing his *Histoire naturelle de l'âme*. But this is not so. Though in *L'Homme machine* he is eager to praise Descartes for his insight into animal nature, he is not, on that account, willing to abandon his earlier view that animals are sentient, intelligent creatures. For shortly after the above remarks, La Mettrie writes:

> Let us observe the ape, the beaver, the elephant, etc., in their operations. If it is clear that these activities can not be performed without intelligence, why refuse intelligence to these animals? And if you grant them a soul, you are lost, you fanatics![36]

So on second thought it might seem that La Mettrie's rhetoric simply got the best of him when he eulogized Descartes. But this is not quite satisfactory either. At the beginning of *Les Animaux plus que machines* he writes:

> Before Descartes no philosopher had regarded animals as machines. Since this famous man only a single and most courageous modern has dared to revive an opinion which seemed condemned to oblivion and even to perpetual scorn; not in order to avenge his compatriot, but carrying temerity to the highest point, in order to apply to man, without any evasion, that which had been said of animals, in order to degrade him, to lower him to that which is most vile, and thus to show no distinction between the master and his subjects.[37]

But the very title of the work indicates that he is not willing to go back on the line of reasoning outlined in II, to

[35] Bussey, pp. 142–43. [36] *Ibid.*, p. 146

[37] *Oeuvres philosophiques*, Vol. II, p. 29.

the effect that animals are sentient and intelligent creatures. Thus, shortly after the above passage, he writes:

> The internal senses are no more lacking to animals than the external ones; consequently they are endowed as we are with the same spiritual faculties which depend on them, I mean perception, memory, imagination, judgment, reason, all things that Boerhaave has proved belong to these senses.[38]

Hence, though textual support may be found for all the claims in both adumbrations I and II, it seems clear that La Mettrie was never really willing to accept Descartes' proof that animals are pure machines (as indicated in I, which obviously is in conflict with other statements there). If this were the case, then to extend that doctrine to man would issue in the claim that nothing thought or felt. This, of course, is absurd, and obviously so. La Mettrie never intended that sort of extension. But it is also clear from the passages cited from *L'Homme machine* and *Les Animaux plus que machines* that at least he thought that Descartes had proved something. The most reasonable interpretation seems to be this: La Mettrie did believe that Descartes had shown animals to be machines, but that he was wrong in thinking (a) that they were *only* machines—devoid of sense and feeling, and (b) that we were anything *more than* sentient, intelligent machines (i.e., possessors of souls). In fact, he is willing to write about Descartes' distinguishing between mind and body as being:

> . . . plainly but a trick of skill, a ruse of style, to make theologians swallow a poison, hidden in the shade of an analogy which strikes everybody else and which they alone fail to notice. For it is this, this strong analogy, which forces all scholars and wise judges to confess that these proud and

[38] *Ibid.*, p. 30.

> vain beings, more distinguished by their pride than by the name of men however much they may wish to exalt themselves, are at bottom only animals and machines which, though upright, go on all fours.[39]

In the *Histoire naturelle de l'âme* there is an anti-Cartesian emphasis, for he wishes to show in the course of that work that Descartes had underestimated animal nature, whereas in *L'Homme machine* he intends to use the Cartesian claim that animals *are* machines to support a mechanistic view of man, since, he believes, we are not essentially different from animals. Thus, La Mettrie does not double back on his initial claim that Descartes was wrong in thinking animals were neither conscious nor intelligent, and this is taken up again in *Les Animaux plus que machines,* though in this work he also wishes to point out that there are great differences *in degree* between our intellectual capabilities and theirs.[40] In short, beginning with *L'Homme machine* La Mettrie expresses great sympathy with Descartes, indicates agreement with his claim that animals are machines, but believes that by showing that there is no essential difference between human beings and animals he can establish both that animals share to some degree our intelligence, while we share to every degree their "machineness." Hence the *bête machine* doctrine comes to be extended to man (*"sans nul detour,"* so he thinks), and thought and feelings are "extended" to animals. But to do this La Mettrie must convincingly show that there are no essential differences between men and animals. It is clear what this involves, an attack on the doctrine that beasts are incapable of learning a language, which Descartes and others thought was possible only for a subject which possessed a soul. This is the challenge which La Mettrie

[39] Bussey, p. 143.

[40] For example, see *Oeuvres philosophiques,* Vol. II, pp. 78–81.

presented to Descartes. *In effect* he accepts Descartes' claim that if and only if S cannot pass test *L* (and *A*), then S does not possess "characteristic" r (capacity to reason, think, etc.) but he denies Descartes' implicit second premise that beasts (and machines) are subjects which cannot pass test *L* (and *A*). This becomes obvious from his discussion of teaching an ape to speak, and the possibility of some day building a talking mechanical man.

After La Mettrie has said that he does not believe the construction of a talking mechanical man would be impossible "especially in the hands of another Prometheus," he goes on to write:

> In like fashion, it was necessary that nature should use more elaborate art in making and sustaining a machine which for a whole century could mark all motions of the heart and of the mind; for though one does not tell time by the pulse, it is at least the barometer of the warmth and the vivacity by which one may estimate the nature of the soul. I am right! The human body is a watch, a large watch constructed with such skill and ingenuity.[41]

The "In like fashion" compares the art of making and sustaining a human body to the less elaborate art which was required of Vaucanson when he constructed his mechanical flute player and mechanical duck.[42] The art required for making the flute player is said to be more elaborate than that needed for making the duck, and that required for making the talking mechanical man is assumed to be more elaborate than that required for the flute player. And the human body is "in like fashion" said to require more elaborate art than that needed for making the flute player, but it is not said to be necessarily more demanding in art than the talking mechanical man. This comparison is not made, and the human body and

[41] Bussey, p. 141. [42] *Ibid.*, pp. 140–41.

the talking mechanical man are placed instead on equal footing relative to the flute player. But even if this were not the case and the human body were said to require more elaborate art than a talking mechanical man, the main point would remain: a human being is said to be nothing over and above a very complex mechanical construction, and the body-machine marks "all motions of the heart and of the mind." Though La Mettrie sometimes talks about man, and sometimes about man's body, it seems clear from *L'Homme machine* that he denies that there is any mental self, soul, or substances over and above that which may be viewed as part of the body itself. As to thought, he writes:

> I believe that thought is so little incompatible with organised matter, that it seems to be one of its properties on a par with electricity, the faculty of motion, impenetrability, extension, etc.[43]

Though he does not delve into the details of the degree of linguistic proficiency which would be exhibited by the talking mechanical man, there is (contra-Lange) no reason why we should not regard La Mettrie as having believed that such an invention would be in some degree rational. The reason for this is simply that La Mettrie does not regard reasoning, thought, the use of language, and the like as anything more than the output of highly complex machines.[44] Furthermore, as we have already

[43] *Ibid.*, pp. 143–44. But see Aram Vartanian's *La Mettrie's L'Homme Machine* for a different interpretation of La Mettrie's position on the mind/body question. Also see his *Diderot and Descartes*, esp. Chap. IV.

[44] Cf. "Nothing, as any one can see, is so simple as the mechanism of our education. Everything may be reduced to sounds or words that pass from the mouth of one through the ears of another into his brain." *Ibid.*, p. 104.

seen, even Descartes never denied the possibility of building a machine so that "if it is touched in a particular part it may ask what we wish to say to it; if in another part it may exclaim that it is being hurt, and so on."[45] It seems likely that the mechanism which La Mettrie claims should no longer be regarded as impossible would be a machine with more linguistic dexterity than the one which Descartes has no trouble imagining. Also, however, La Mettrie in effect claims that the sort of machine which Descartes regarded as possible, is not really so very different from us after all.

Furthermore, it is clear in La Mettrie's discussion of the ape, that he does not envision it as capable only of mere mimicry (or perhaps we should say, mere low-level mimicry). He knew that parrots could do that,[46] and surely knew that Descartes and others had been and were aware of instances of simple mimicry and imitation. He does not seem to be offering just another example of that kind. He envisions teaching an ape a language, not simply to reproduce a few sounds. He talks about taking great care in selecting an intelligent-looking ape, and then seeing if the then famous teacher of deaf mutes–Johann Conrad Amman–would be able to educate it to use a language. Toward the end of the discussion he writes:

> Could not the device which opens the Eustachian canal of the deaf, open that of apes? Might not a happy desire to imitate the master's pronunciation, liberate the organs of speech in animals that imitate so many other signs with

[45] *Discourse,* p. 116.

[46] And also mentions one which could supposedly do more than that: "which could answer rationally, and which had learned to carry on a kind of connected conversation, as we do," Bussey, pp. 101–02. This is mentioned during the course of his discussion of the ape, and is obviously closer to the goals he envisions for the ape than mere mimicry would be.

such skill and intelligence? Not only do I defy anyone to name any really conclusive experiment which proves my view impossible and absurd; but such is the likeness of the structure and functions of the ape to ours that I have very little doubt that if this animal were properly trained he might at last be taught to pronounce, and consequently to know, a language. Then he would no longer be a wild man, nor a defective man, but he would be a perfect man, a little gentleman, with as much matter or muscle as we have, for thinking and profiting by his education.[47]

And the main point made immediately following the above passage is:

The transition from animals to man is not violent, as true philosophers will admit. What was man before the invention of words and the knowledge of language? An animal of his own species with much less instinct than the others.[48]

Thus there can be no doubt that La Mettrie envisioned, in effect, animals (machines) which could pass Descartes' language test. Furthermore, there can be no doubt that La Mettrie was aware of the test and intent on impugning the distinction it was devised to uphold. Sometimes he seems to be saying that the test simply isn't a fair one (because animals have their own sort of language), and that mimicry is not so very different from the ways in which we learn to speak. Thus in the *Histoire naturelle de l'âme,* he writes:

Conventional language, I mean speech, is not the sign which expresses it best. There is another one common to men and animals which shows it with greater certainty; I speak of emotional language, such as laments, cries, caresses, evasion, sighs, song, and in a word, all the expressions of pain, sadness, aversion, fear, audacity, submission, anger, pleasure, joy, tenderness, etc. Such an energetic language

[47] *Ibid.,* pp. 102–03. [48] *Ibid.,* p. 103.

has much greater power to convince us than have all the sophisms of Descartes.[49]

And:

> But do men and even women make fun of each other better than do the birds who repeat the songs of other birds in such a way as to ridicule them perfectly? What difference is there between the infant and the parrot that is trained: Don't they equally repeat the sounds which strike their ears, and each do it with as little intelligence as the other?[50]

And later in *Les Animaux plus que machines* he indicates both that he does not think so highly of the test, but that even so, it's not impossible that animals should pass it:

> But, it is said, animals do not have speech! Wonderful objection! Say also that they walk on four feet and see the sky only when sleeping on their backs, and finally reproach the author of nature for the innocent pleasure that he has taken in varying his works.
>
> Who deprives animals of the gift of speech? *Nothing,* perhaps. This *nothing* of Fontenelle distinguishes him as much from all other men as men are distinguished from brutes. Perhaps one day this weak obstacle will be removed; the thing isn't impossible, according to the author of *l'homme machine.* The enticing example of his great ape, and the wonderful projects which have passed through his head.[51]

No doubt one of the wonderful projects which would help to raise the obstacle referred to, was none other than the talking mechanical man.

Thus it is evident that La Mettrie was at least in a general way aware of Descartes' arguments for making a hard and fast distinction between man and beast, and

[49] *Oeuvres philosophiques,* Vol. I, p. 84.
[50] *Ibid.,* p. 152.
[51] *Oeuvres philosophiques,* Vol. II, pp. 31–32.

wishes to challenge them. But it is also evident that by the time he wrote *L'Homme machine* he was willing to praise Descartes' *bête machine* doctrine, and fancy himself as merely extending that doctrine to man. It is these two facts which unveil the very imperfect understanding which La Mettrie had of his predecessor's mechanism. Because to argue that an animal (or machine) could pass the language test, was in effect to vitiate the way in which Descartes thought he was able to prove that animals were machines. La Mettrie never understood that the language test was an essential part of Descartes' way of arguing to the conclusion that La Mettrie was in sympathy with, the claim that animals are machines. It is as if La Mettrie said, "Now Descartes has shown us that animals are machines; I accept his proof, but he hasn't shown us that these machines can't think or feel; he thinks the use of language distinguishes us who are thinking feeling creatures from these machines, but I believe I can show that animals also have linguistic capabilities; hence we find that they are machines which think and feel, and, since they are like us and we like them, we too are machines." But this is really to show (assuming the argument is sound) that Descartes hasn't proved anything at all. For Descartes' arguments could only show that beasts were *pure* machines. And they could only do so if they could be based on something like the language test. So La Mettrie cannot really accept any part of the *bête machine* aspect of Descartes' mechanism and remain consistent with his other claims. He does not clearly see that his modified "impure" *bête machine* doctrine cannot make use of the same arguments Descartes used to establish his "pure" *bête machine* doctrine. He believed that since Descartes claimed that language separated us from animals and machines, to show that animals and machines could learn a language would be to show that we were nothing but machines which could think and feel. But

Descartes would never accept that conclusion. For him it was a contradiction to suppose that machines could think or feel. Thus to get an animal, let us say, to pass the language test would not persuade him that we were only machines, it would rather simply force him to admit that he had not established the absence of souls in animals, and that he had thus provided no solution to the theological problems which the *bête machine* doctrine was in part designed to solve. To admit that we are only machines whose complexity and arrangement of parts give rise to thought and consciousness would be in effect to deny the soul. It seems unlikely that anything could have persuaded Descartes to do that. Even if "another Prometheus" made a highly convincing talking mechanical man, I believe it is more likely that Descartes would rather have claimed that a generous God had granted the clever fellow an extra soul to go with his invention, than submit to the conclusion that we had no soul at all.

And it is also worth noting that La Mettrie never fully grasped other specifics of the *bête machine* doctrine either. If he had realized that the way in which Descartes developed his argument involved the assumption that there is a physiological parallel between man's vocal faculties and those of the animals, he would have understood that to attribute the disparity in linguistic ability between man and beast to slight physical variations involved challenging the empirical assumption Descartes needed to establish the need for a non-mechanistic explanation of human linguistic performances.

V

There is, of course, a continuity between Descartes' *bête machine* doctrine and the *homme machine* of La Mettrie. But it is important to point out the kind of continuity that it is, namely, one that involves the inconsistent adaptations of half-understood views of one's predecessor. It is this that is sometimes overlooked by commentators who speak of the dangers of a thoroughgoing mechanism that were ever-present in the Cartesian *bête machine* doctrine. For example, Balz has written:

> The more successfully physiology and neurology could be worked out on the dualistic, that is, on the mechanistic basis, the more facts that could reasonably be explained as due to the machine, the less the need for an animal soul. But success in dispensing with an animal soul brought precisely the danger that the argument would be extended to man, to the undermining of his privileged position. If the animal can satisfactorily be described as a machine, then the more varied and the greater the powers of the animal, the greater the probability of the inference that man too is an automaton.[52]

But it is not indicated that Descartes dispensed with the animal soul via the *bête machine* doctrine (not by his physiologizing which applied equally to man and animals), and that the *bête machine* doctrine implied, within Descartes' framework, the preservation of a strict distinction between man and automata.

Yet there is an understandable reason why this implication is not always clearly seen. This is simply because the *bête machine* doctrine and what we referred to earlier as the "physiologizing of the soul" are often described (easily, but indiscriminately) as Cartesian mechanism (as in Balz's remarks above.) They are, to be sure, both parts of Cartesian mechanism, but they nevertheless involve very different assumptions and arguments. Both "kinds"

[52] *Cartesian Studies,* p. 110. For similar construals of the continuity between the *bête machine* doctrine and the *homme machine,* see E. G. Boring's *A History of Experimental Psychology,* p. 623; J. P. Mahaffy's *Descartes,* pp. 63–64; and A. Vartanian's *Diderot and Descartes,* esp. Chap. 5.

of mechanism are present in La Mettrie's writings, but the *bête machine* doctrine as propounded by Descartes stands, in every essential, unalterably opposed to a thoroughgoing mechanism. It is dualistic since physiological parallels between man and beast coupled with non-parallels in linguistic performance are interpreted as requiring a non-physiological account of human verbal performances. On the other hand, what I have called the "physiologizing of the soul" did provide La Mettrie with a more logically consistent source for his thoroughgoing mechanism. As Lange points out:

> A mechanism of pressure and collision, which Descartes follows out with great ingenuity through all the separate steps, forms an uninterrupted chain of effects produced by external things through the senses upon the brain, and from the brain back again outwards through nerves and muscular filaments. . . .
>
> In this state of things we may seriously ask whether De la Mettrie was not in truth quite justified when he traced his own Materialism to Descartes, and maintained that the wily philosopher, purely for the sake of the parsons, had patched on to his theory a soul, which was in reality quite superfluous.[53]

And here, I believe, is the real and only "danger" which Cartesianism posed for the churchmen. At any rate, those who thought a thoroughgoing mechanism was also possible via the *bête machine* doctrine failed to understand the arguments which supported it. Here, of course (contra-Lange), the soul was not superfluous.

Father Malebranche in "The Defence of the Author of the Treatise concerning the Search After Truth against the Accusation of Monsieur de la Ville" writes:

> Let a Man but say in Company, with an Air of Gravity, or rather with a Look into which the Imagination, scar'd

[53] *History of Materialism,* p. 246.

> with something extraordinary, forms the Face: *Really the Cartesians are strange People; They maintain, That Beasts have no Soul. I am afraid in a little time they will say as much of Man:* And this will be enough to persuade a great many that this is a dangerous Opinion.[54]

Malebranche, in keeping with Descartes' own views, was chiding certain worriers for their fears and hesitations in accepting Cartesianism because of the thoroughgoing mechanism which many thought to be tucked between the propositions which made up the *bête machine* doctrine. Malebranche had good reasons for chiding them. Their fears of that doctrine rested on a misunderstanding (just as La Mettrie's praise of that doctrine rested on a misunderstanding). It would be rather like some imaginary defender of Ryle some day having to say:

> Let a Man but say in seminar, with an air of seriousness, or rather with a look of startled apprehension: *Really the Ryleians are strange People; They maintain, that the Mind is not some Thing or Process. I am afraid that in a little time they will say as much of the Brain and Conduction at the Synapse.* And this will be enough to persuade a great many that this is a dangerous opinion.

Absurd to imagine philosophers misinterpretating Ryle to *that* extent. Yet this depicts a situation not so unlike that which confronted those who like Malebranche seemingly understood the details of the *bête machine* doctrine. The above-mentioned fears, of course, were in a sense justified, for there is nothing to guarantee that a doctrine will not be misused by those who hear of it. It was only "a little time" (thirty-two years) after Malebranche's death, that La Mettrie did use the doctrine in order to "say as much of Man." But at least he did not do so without arguing inconsistently. Whether or not there is some way

[54] Trans. by T. Taylor, incl. in same vol. as the *Treatise Concerning the Search After Truth,* cited above, p. 201.

to argue without contradiction for a thoroughgoing mechanism is still at the heart of recent controversies on mentality and machines. Cyberneticians and those working in the areas of computer simulation of cognitive processes and artificial intelligence may turn out to be a kind of collective "Prometheus" who will provide us with the linguistically proficient mechanical man envisioned by La Mettrie.

Current barriers to such an achievement are discussed in Chapter Four, whereas in the chapter immediately following I raise Cartesian-type objections to a certain amiable "animism" that underlies various modern assessments of the potential mental prowess of computers.

CHAPTER TWO

The Imitation Game

I

Disturbed by what he took to be the ambiguous, if not meaningless, character of the question "Can machines think?," the late A. M. Turing in his article "Computing Machinery and Intelligence" sought to replace that question in the following way. He said:

> The new form of the problem can be described in terms of a game which we call the "imitation game." It is played with three people, a man (A), a woman (B), and an interrogator (C) who may be either sex. The interrogator stays in a room apart from the other two. The object of the game for the interrogator is to determine which of the other two is the man and which is the woman. He knows them by labels X and Y, and at the end of the game he says either "X is A and Y is B" or "X is B and Y is A." The interrogator is allowed to put questions to A and B thus:
>
> C: "Will X please tell me the length of his or her hair?"
>
> Now suppose X is actually A, then A must answer. It is A's object in the game to try to cause C to make the wrong identification. His answer might therefore be
>
> "My hair is shingled, and the longest strands are about nine inches long."
>
> In order that tones of voice may not help the interrogator the answers should be written, or better still, typewritten. The ideal arrangement is to have a teleprinter communicating between the two rooms. Alternatively the question and answers can be repeated by an intermediary. The object

> of the game for the third player (B) is to help the interrogator. The best strategy for her is probably to give truthful answers. She can add such things as "I am the woman, don't listen to him!" to her answers, but it will avail nothing as the man can make similar remarks.
>
> We now ask the question, "What will happen when a machine takes the part of A in this game?" Will the interrogator decide wrongly as often as when the game is played between a man and a woman? These questions replace our original, "Can machines think?"

And Turing's answers to these latter questions are more or less summed up in the following passage: "I believe that in fifty years' time it will be possible to program computers, with a storage capacity of about 10^9, to make them play the imitation game so well that an average interrogator will not have more than 70 per cent chance of making the right identification after five minutes of questioning." And though he goes on to reiterate that he suspects that the original question "Can machines think?" is meaningless, and that it should be disposed of and replaced by a more precise formulation of the problems involved (a formulation such as a set of questions about the imitation game and machine capacities), what finally emerges is that Turing does answer the "meaningless" question after all, and that his answer is in the affirmative and follows from his conclusions concerning the capabilities of machines which might be successfully substituted for people in the imitation-game context.

It should be pointed out that Turing's beliefs about the possible capabilities and capacities of machines are not limited to such activities as playing the imitation game as successfully as human beings. He does not, for example, deny that it might be possible to develop a machine which would relish the taste of strawberries and cream, though he thinks it would be "idiotic" to attempt to make one, and confines himself on the whole in his positive account to considerations of machine capacities which could be illustrated in terms of playing the imitation game.

So we shall be primarily concerned with asking whether or not a machine, which could play the imitation game as well as Turing thought it might, would thus be a machine which we would have good reasons for saying was capable of thought and what would be involved in saying this.

Some philosophers have not been satisfied with Turing's treatment of the question "Can machines think?"[1] But the imitation game itself, which indeed seems to constitute the hub of his positive treatment, has been little more than alluded to or remarked on in passing. I shall try to develop in a somewhat more detailed way certain objections to it, objections which, I believe, Turing altogether fails to anticipate. My remarks shall thus in the main be critically oriented, which is not meant to suggest that I believe there are no plausible lines of defense open to a supporter of Turing. I shall, to the contrary, close with a brief attempt to indicate what some of these might be and some general challenges which I think Turing has raised for the philosopher of mind.

II

Let us consider the following question: "Can rocks imitate?" One might say that it is a question "too mean-

[1] See Michael Scriven, "The Mechanical Concept of Mind," pp. 31ff., and "The Compleat Robot: A Prolegomena to Androidology" in *Dimensions of Mind,* ed. Sidney Hook (New York: New York University Press, 1960). Also a remark by Paul Ziff in "The Feelings of Robots," pp. 98ff., and others—for example, C. E. Shannon and J. McCarthy in their preface to *Automata Studies* (Princeton: Princeton University Press, 1956). Ziff's over-all position is discussed in the next chapter.

ingless to deserve discussion." Yet it seems possible to reformulate the problem in relatively unambiguous words as follows:

> The new form of the problem can be described in terms of a game which we call the "toe-stepping game." It is played with three people, a man (A), a woman (B), and an interrogator (C) who may be of either sex. The interrogator stays in a room apart from the other two. The door is closed, but there is a small opening in the wall next to the floor through which he can place most of his foot. When he does so, one of the other two may step on his toe. The object of the game for the interrogator is to determine, by the way in which his toe is stepped on, which of the other two is the man and which is the woman. He knows them by labels X and Y, and at the end of the game he says either "X is A and Y is B" or "X is B and Y is A." Now the interrogator–rather the person whose toe gets stepped on–may indicate before he puts his foot through the opening, whether X or Y is to step on it. Better yet, there might be a narrow division in the opening, one side for X and one for Y (one for A and one for B).
>
> Now suppose C puts his foot through A's side of the opening (which may be labeled X or Y on C's side of the wall). It is A's object in the game to try to cause C to make the wrong identification. His step on the toe might therefore be quick and jabbing like some high-heeled woman.
>
> The object of the game for the third player (B) is to help the person whose toe gets stepped on. The best strategy for her is probably to try to step on it in the most womanly way possible. She can add such things as a slight twist of a high heel to her stepping, but it will avail nothing as the man can step in similar ways, since he will also have at his disposal various shoes with which to vary his toe-stepping.
>
> We now ask the question: "What will happen when a rock box (a box filled with rocks of varying weights, sizes, and shapes) is constructed with an electric eye which oper-

ates across the opening in the wall so that it releases a rock which descends upon C's toe whenever C puts his foot through A's side of the opening, and thus comes to take the part of A in this game?" (The situation can be made more convincing by constructing the rock box so that there is a mechanism pulling up the released rock shortly after its descent, thus avoiding telltale noises such as a rock rolling on the floor, etc.) Will then the interrogator–the person whose toe gets stepped on–decide wrongly as often as when the game is played between a man and a woman? These questions replace our original, "Can rocks imitate?"

I believe that in less than fifty years' time it will be possible to set up elaborately constructed rock boxes, with large rock-storage capacities, so that they will play the toe-stepping game so well that the average person who would get his toe stepped on would not have more than 70 per cent chance of making the right identification after about five minutes of toe-stepping.

The above seems to show the following: what follows from the toe-stepping game situation surely is not that rocks are able to imitate (I assume no one would want to take that path of argument) but only that they are able to be rigged in such a way that they could be substituted for a human being in a toe-stepping game without changing any essential characteristics of that game. And this is claimed in spite of the fact that if a human being were to play the toe-stepping game as envisaged above, we would no doubt be correct in saying that that person was imitating, etc. To be sure, a digital computer is a more august mechanism than a rock box, but Turing has not provided us with any arguments for believing that its role in the imitation game, as distinct from the net results it yields, is any closer a match for a human being executing such a role, than is the rock box's execution of its role in the toe-stepping game a match for a human being's execution of a similar role. The parody comparison can be pushed too far. But I think it lays bare the reason why there is no contradiction involved in saying, "Yes, a machine can play the imitation game, but it can't think." It is for the same reason that there is no contradiction in saying, "Of course a rock box of such-and-such a sort can be set up, but rocks surely can't imitate." For thinking (or imitating) cannot be fully described simply by pointing to net results such as those illustrated above. For if this were not the case it would be correct to say that a phonograph could sing, and that an electric eye could see people coming.

People may be let out of a building by either an electric eye or a doorman. The end result is the same. But though a doorman may be rude or polite, the electric eye neither practices nor neglects etiquette. Turing brandishes net results. But I think the foregoing at least indicates certain difficulties with any account of thinking or decision as to whether a certain thing is capable of thought which is based primarily on net results. And, of course, one could always ask whether the net results were really the same. But I do not wish to follow that line of argument here. It is my main concern simply to indicate where Turing's account, which is cast largely in terms of net results, fails because of this. It is not an effective counter to reply: "But part of the net results in question includes intelligent people being deceived!" For what would this add to the general argument? No doubt people could be deceived by rock boxes! It is said that high-fidelity phonographs have been perfected to the point where blindfolded music critics are unable to distinguish their "playing" from that of, let us say, the Budapest String Quartet. But the phonograph would never be said to have performed with unusual brilliance on Saturday, nor would it ever deserve an encore.

III

Now perhaps comparable net results achieved by machines and human beings is all that is needed to establish an analogy between them, but it is far from what is needed to establish that one sort of subject (machines) can do the same thing that another sort of subject (human beings or other animals) can do. Part of what things do is how they do it. To ask whether a machine can think is in part to ask whether machines can do things in certain ways.

The above is relevant to what might be called the problem of distinguishing and evaluating the net results achieved by a machine as it is touched on by Scriven in his discussion of what he calls "the performatory problem" and "the personality problem." In "The Compleat Robot: A Prolegomena to Androidology," he writes:

> The performatory problem here is whether a computer can produce results which, when translated, provide what would count as an original solution or proof *if it came from a man.* The personality problem is whether we are entitled to call such a result a solution or proof, despite the fact that it did *not* come from a man.

And continues:

> The logical trap is this: no *one* performatory achievement will be enough to persuade us to apply the human-achievement vocabulary, but if we refuse to use this vocabularly in each case separately, on this ground, we will, perhaps wrongly, have committed ourselves to avoiding it even when *all* the achievements are simultaneously attained.[2]

[2] In *Dimensions of Mind,* ed. Sidney Hook, pp. 118–42.

My concern is not, however, with what is to count as an original solution or proof. Scriven, in the above, is commenting on the claims: "Machines only do what we tell them to do. They are incapable of genuinely original thought." He says that two "importantly different points are run together." The above is his attempt to separate these points. But it seems that there are at least three, and not just two, points which are run together in the just-mentioned claims. The third point, the one not covered by Scriven's distinction between the performatory[3] and personality problems, is simply the problem, mentioned above, of discerning when one subject (a machine) has *done the same thing* as another subject (a human being). And here "doing the same thing" does not simply mean "achieved similar end result." (Which is not to suggest that the phrase can never be used in that way in connection with thinking.) This is of interest in respect to Scriven's discussion, since it might be the case that all the achievements were simultaneously attained by a machine, as Scriven suggests, and that we had decided on various grounds that they should count as original proofs and solutions and thus surmounted the personality problem, but yet felt unwilling to grant that the machines were capable of "genuinely original thought." Our grounds for this latter decision might be highly parallel to our grounds for not wanting to say that rocks could imitate (even though rock boxes had reached a high level of development). Of course our grounds might not be as sound as these. I am simply imagining the case where they are, which is also a case where all the achievements are attained in such a way that they count as original solutions or proofs. In this case we would see that answers to questions about originality and performance and the

[3] In spite of my remarks on originality, which follow, Scriven's general point seems very well taken.

logical trap mentioned by Scriven would be wholly separate from whatever answers might be given to the question whether or not the machines involved thought, and would thus be unsuitable as answers to the question whether or not they were capable of "genuinely original thought." In other words, questions as to originality and questions as to thinking are not the same, but this dissimilarity is left unacknowledged in Scriven's account.

IV

Suppose we build a machine that X-es, where X-ing is arriving at a certain result or conclusion (which may also be referred to as an activity which may or may not involve the use of language) and which is denoted by verbs or verb phrases such as "calculates" or "computes" or "utters words" or "finds its way home in the spring" or, as in Descartes' clock example, "measures time." Let us assume that the results denoted by such verbs may be graded in various ways, so that the subjects responsible for them may be said to have performed "better than average," "in a superior way," "as well as," "somewhat better than," and so forth, relative to the performances of other subjects which could be human beings, Saturnians, animals, machines, *et al.* Though we shall sometimes appear to be talking only about the skills or abilities of an individual, we shall always wish it understood that the individual represents a certain subset of subjects, sometimes, for example, a "species" of subjects (animals), sometimes only a kind of member of a species or a kind of man–such as physicist, artist, *et al.* Now someone might argue on behalf of a machine, and in the same fashion that Montaigne argued on behalf of beasts, that if it is able to X "as well as" or "in a superior way to" the X-ing of which a man is capable, then, if man's X-ing would require that he thinks, reasons, acts intelligently, and so forth, so does the machine's X-ing, and its ability in doing so is indicated by the way in which we would grade it. Descartes, we have already emphasized, was intent on denying (and rightly so, I believe) the correctness of this sort of reasoning. For if the case where the machine X-es is really the same and not just vaguely analogous to the case where the man X-es, then we should be safe in making certain further assumptions about the machine's general capabilities and performances, just as in the case where we know that a man can do X and must thus be able to do a number of other things as well. For example, if a man does X where X-ing is devising a theory for splitting the atom, then we may generally assume that such a man is capable of a great many other things as well (we may assume that scientific theorists are capable of a great many other things as well); we may assume that he knows some mathematics and physics, that he has learned a language, that he can solve a wide variety of problems, and so on. Some of these skills would be closely related to devising such theories, some would be less closely related. Now, though we would not be safe in assuming that our atomic scientist would be a crackerjack poet or tennis player, we would be safe in assuming that he would be capable of some cluster of abilities such as those just-mentioned and that we can expect further things from him, that he might perhaps be just the man to teach in a physics department at some university, would be a good man to review other works in the field, should be a first-rate adviser to the Atomic Energy Commission, and so on. I believe the above preserves the core of what Descartes was claiming when he wrote about certain performances of beasts that "the fact that they do better than we do, does not prove that they are endowed with mind, for in this case they would have more reason than any of us, and would surpass us in all other things."[4] (And, of course, it is assumed that they don't surpass us in all—or perhaps even any—other things.) I believe that the phrase "in many other" or even "in various other" or "in a wide range of" could be substituted for the phrase "in all other" without in any way changing the main point of what Descartes was trying to establish with the above statement.

Take the case of a machine again, where that machine is, let us say, a digital computer loaded with a certain program so that in a matter of minutes it figures out monthly checks of all the workers at General Motors. Now if any human being were able to do that, we should be safe in assuming that, since for most of us it would require a great deal of mental effort, figuring, and cross-checking, the very efficient human being would also outdo us at all sorts of other tasks, would be better at all sorts of monetary estimations, mathematical puzzles, and so on. Or, better still, take the case of a swallow which returns in the spring to its place of nesting. If a man were, let us say, able to drive from Princeton to Capistrano after merely glancing at a road map, we should be able to assume that he's good at maps, just the fellow to tell us how to get out of downtown Manhattan, or, if he can't verbalize it, just the man to drive us out, and so forth. Or suppose a machine is set up to slosh some paint around on a canvas, that someone enters it in an art contest, and that the machine-painting wins the prize. If a man has a painting of his submitted in a reputable contest and wins a prize, we may generally assume he's a well-trained painter, an artist with a number of attending skills. So then, a Montaigne or a Turing might say, is the ma-

[4] *Loc. cit.*

chine not an artist? He just took the prize from Jackson Pollock, Arshile Gorky, and Willem de Kooning. No, Descartes is saying. Just as in the earlier case of the digital computer where we were not entitled to assume that it could do anything other than what it was programmed to do[5]—namely, figure out wages—and in the case of the swallows we can't assume that they're good at maps, terrain, landmarks, or anything else in general, so too in the case of the art contest we can't assume that such a machine had in some way worked out sketches leading up to the prize painting, that it had experimented with various textures, colorings, techniques, or that it had some knack with prints, water colors, or was capable of *anything* other than being able to slosh paint about in the way it did. We can't assume that it has any artistic habits, skills, or techniques at all. Its art has not developed and will not develop. Similarly the case of the digital computer is not like the case of a man who has acquired a number of general computational skills and applies them to a variety of problems and situations. It is instead like the case of a man who was only able to do a speedy computation of salaries for General Motors' workers for a particular month. The case of the swallow is not like the case of the fellow who is good at maps, but would be like the case of a fellow who was only able to get from Princeton to Capistrano to Princeton to . . . , etc. In other words, as in the case of certain computer outputs

[5] But I am not yet dealing with the general question of what, in principle, a computer might be programmed to do. Nor am I assuming that certain types of computer programs would not be more flexible from situation to situation, would not have to be replaced each time, would not be self-adjusting in various ways, etc. I am simply discussing the case where a new situation *does* demand a wholly new program, and the computer's operation issues in a certain single result, (Or, in Descartes' words, whose "organs have need of some special adaptation for every particular action.")

—a poem, for example[6]—we have hitherto understood the result in question to be such that its production required certain general skills or capacities on the part of human beings. And human beings who possessed such general skills or capacities could be safely assumed to be able to do a number of other things too. Hence if the machine truly writes poems in at least roughly the way that a human being writes poems—which would be the only sort of case where we would be justified in assuming that it was able to understand a language, reason, and reflect—then we should also be able to assume that the machine is also capable of a wide range of other activities, in which verbal, thinking, reasoning, and reflecting creatures are capable of participating. But in the case of the computer producing a poem (or a parrot, words) we cannot correctly make such an assumption. (Which, of course, does not mean that no poem has been written, nor words uttered.) Hence why not look for an alternative explanation? This is precisely what Descartes did.

Put in another way, Descartes pointed out in effect the simple but important (and often ignored) fact that we do not understand a man doing a calculation, for example, simply by observing that one isolated case of his calculating. We do not understand, nor should we assess or describe, his performance in a particular case apart from all the other things we have learned about him and men in general. There is a whole network of knowledge which we bring to his situation, and it is in large part on the basis of

[6] As good or better than the one which the LGP-30 under the direction of R. F. Reiss and R. M. Worth composed called *Ode to a Depot:*

> By the new neighbors their depot was jade,
> These ulcers were new, many depots were suede.

In *San Francisco Review,* Vol. I, No. 12, June 1962, p. 86. The computer's *Elegy for a Lady* is closer to the style of Dylan Thomas; see p. 85.

this that we are willing and able to say what we say about him, that we say here is a man who is struggling with a problem, thinking things out, checking his results, and so forth. If it were another sort of man whom we watched the diabolical neurosurgeon rig up so that all he would do hour after hour was calculate the monthly wages of various firms, then we should be expected to appraise his activity accordingly; we should say, "Well, I know he seems at first to be a super estimator, but you see this neurosurgeon rigged his brain, and so on, the poor devil's optic nerves have been fixed so that he only sees printed names and figures, can't really find his own way to the door." Here we would have an alternative explanation of how results usually obtained by thinking and reflection were brought about in another way. It is only reasonable to expect the explanation to be used—so too in the case of a computer computing.

A rock rolls down a hill and there is, strictly speaking, no behavior or action on the part of the rock. But if a man rolls down a hill we might well ask if he was pushed or did it intentionally, whether he's enjoying himself, playing a game, pretending to be a tumbleweed, or what. We cannot think of a man as simply or purely rolling down a hill—unless he is dead. *A fortiori,* we cannot understand him being a participant in the imitation game apart from his dispositions, habits, etc., which are exhibited in contexts other than the imitation game. Thus we cannot hope to find any decisive answer to the question as to how we should characterize a machine which can play (well) the imitation game, by asking what we would say about a man who could play (well) the imitation game. Thinking, whatever positive characterization or account is correct, is not something which any one example will explain or decide. But the part of Turing's case which I've been concerned with rests largely on one example.

V

The following might help to clarify the above. Imagine the dialogue below:

Vacuum Cleaner Salesman: Now here's an example of what the all-purpose Swish 600 can do. (He then applies the nozzle to the carpet and it sucks up a bit of dust.)

Housewife: What else can it do?

Vacuum Cleaner Salesman: What do you mean "What else can it do?" It just sucked up that bit of dust, didn't you see?

Housewife: Yes, I saw it suck up a bit of dust, but I thought it was all-purpose. Doesn't it suck up larger and heavier bits of straw or paper or mud? And can't it get in the tight corners? Doesn't it have other nozzles? What about the cat hair on the couch?

Vacuum Cleaner Salesman: It sucks up bits of dust. That's what vacuum cleaners are for.

Housewife: Oh, that's what it does. I thought it was simply an example of what it does.

Vacuum Cleaner Salesman: It is an example of what it does. What it does is to suck up bits of dust.

We ask: Who's right about examples? We answer: It's not perfectly clear that anyone is lying or unjustifiably using the word "example." And there's no obvious linguistic rule or regularity to point to which tells us that if S can only do X, then S's doing X cannot be an example of what S can do since being an example presupposes or entails or what not that other kinds of examples are forthcoming (sucking up mud, cat hair, etc.). Yet, in spite of this, the housewife has a point. One simply has a right to expect more from an all-purpose Swish 600 than what has been demonstrated. Here clearly the main trouble is with "all-purpose" rather than with "example," though there may still be something misleading about saying, "Here's an example . . . ," and it would surely mislead to say, "Here's *just* an example . . . ," followed by ". . . of what the all-purpose Swish 600 can do." The philosophical relevance of all this to our own discussion can be put in the following rather domestic way: "thinking" is a term which shares certain features with "all-purpose" as it occurs in the phrase "all-purpose Swish 600." It is not used to designate or refer to one capability, capacity, disposition, talent, habit, or feature of a given subject any more than "all-purpose" in the above example is used to mark out one particular operation of a vacuum cleaner. Thinking, whatever positive account one might give of it, is not, for example, like swimming or tennis playing. The question as to whether Peterson can swim or play tennis can be settled by a few token examples of Peterson swimming or playing tennis. (And it might be noted it is hardly imaginable that the question as to whether Peterson could think or not would be raised. For in general it is not at all interesting to ask that question of contemporary human beings, though it might be interesting for contemporary human beings to raise it in connection with different anthropoids viewed at various stages of their evolution.) But if we suppose the question were raised in connection with Peterson, the only appropriate sort of answer to it would be one like, "Good heavens, what makes you think he can't?" (as if anticipating news of some horrible brain injury inflicted on Peterson). And our shock would not be at his perhaps having lost a particular talent. It would not be like the case of a Wimbledon champion losing his tennis talent because of an amputated arm.

It is no more unusual for a human being to be capable of thought than it is for a human being to be composed of cells. Similarly, "He can think" is no more an answer to questions concerning Peterson's mental capacities or intelligence, than "He's composed of cells" is an answer to the usual type of question about Peterson's appearance. And to say that Peterson can think is not to say there are a few token examples of thinking which are at our fingertips, any more than to say that the Swish 600 is all-purpose is to have in mind a particular maneuver or two of which the device is capable. It is because thinking cannot be identified with what can be shown by any one example or type of example; thus Turing's approach to the question "Can a machine think?" via the imitation game is less than convincing. In effect he provides us below with a dialogue very much like the one above:

Turing: You know, machines can think.
Philosopher: Good heavens! Really? How do you know?
Turing: Well, they can play what's called the imitation game. (This is followed by a description of same.)
Philosopher: Interesting. What else can they do? They must be capable of a great deal if they can really think.
Turing: What do you mean, "What else can they do?" They play the imitation game. That's thinking, isn't it?
Etc.

But Turing, like the vacuum cleaner salesman, has trouble making his sale. Nonetheless, I will indicate shortly why certain of our criticisms of his approach might have to be modified.

VI

But one last critical remark before pointing to certain shortcomings of the foregoing. As indicated before, Turing's argument benefits from his emphasizing the fact that a machine is being substituted for a human being in a certain situation, and does as well as a human being would do in that situation. No one, however, would want to deny that machines are able to do a number of things as well as or more competently than human beings, though surely no one would want to say that every one of such examples provided further arguments in support of the claim that machines can think. For in many such cases one might, instead of emphasizing that a machine can do what a human being can do, emphasize that one hardly needs to be a human being to do such things. For example: "I don't even have to think at my job; I just seal the jars as they move along the belt," or, "I just pour out soft drinks one after the other like some machine." The latter could hardly be construed as suggesting "My, aren't soft-drink vending machines clever," but rather suggests, "Isn't my job stupid; it involves little or no mental effort at all." Furthermore, as Professor Ryle has suggested to me, a well-trained bank cashier can add, subtract, multiply, and divide without having to think about what he is doing and while thinking about something else, and can't many of us run through the alphabet or a popular song without thinking? This is not meant to be a specific criticism of Turing as much as it is meant as a reminder that being able to do what human beings can do hardly implies the presence of intellectual or mental skills real or simulated, since so many things which human beings do involve little, if any, thinking. Those without jobs constitute a somewhat different segment of the population from those without wits.

VII

But the following considerations seem to temper some of the foregoing criticisms. A defender of Turing might emphasize that a machine that is able to play the imitation game is also able to do much more; it can compute, perhaps be programmed to play chess, etc., and consequently displays capacities far beyond the "one example" which has been emphasized in our criticisms. I shall not go into the details which I think an adequate reply to this challenge must take into account. But in general I believe it would be possible to formulate a reply along the lines that would show that even playing chess, calculating, and the performance of other (most likely computational) operations provide us with at best a rather narrow range of examples and still fails to satisfy our intuitive concept of thinking. The parallel case in respect to the Housewife and Vacuum Cleaner Salesman would be where the Housewife still refused to accept the vacuum cleaner as "all-purpose" even though it had been shown to be capable of picking up scraps somewhat heavier than dust. Nonetheless, even if our reply were satisfactory, the more general question would remain unanswered: what range of examples would satisfy the implicit criteria we use in our ordinary characterization of subjects as "those capable of thought"?

A corollary: If we are to keep the question "Can machines think?" interesting, we cannot withhold a positive answer simply on the grounds that it (a machine) does not duplicate human activity in every respect. The question "Can a machine think if it can do everything a human being can do?" is not an interesting question, though it is at least interesting that some philosophers have thought it interesting to ask whether there would not be a logical contradiction in supposing such to be, in fact, a machine. But as long as we have in mind subjects which obviously are machines, we must be willing to stop short of demanding their activities to fully mirror human ones before we say they can think, if they can. But how far short? Again the above question as to the variety and extent of examples required is raised.

Furthermore, it might be asserted that with the increasing role of machines in society the word "think" itself might take on new meanings, and that it is not unreasonable to suppose it changing in meaning in such a way that fifty years hence a machine which could play the imitation game would in ordinary parlance be called a machine which could think. There is, however, a difference between asking whether a machine can think given current meanings and uses of "machine" and "think" and asking whether a machine can think given changes in the meanings of "machine" and "think." My own attention has throughout this chapter centered on the first question. Yet there is a temporal obscurity in the question "Can machines think?" For if the question is construed as ranging over possible futures, it may be difficult to discuss such futures without reference to changing word uses and senses. To some extent Turing's own views are based on certain beliefs he has about how we will talk about machines in the future. But these are never discussed in any detail, and he does not address himself to the knotty problems of meaning which interlace with them. Some of these issues are discussed in section VI of the next chapter.

VIII

The stance is often taken that thinking is the crowning capacity or achievement of the human race, and that if one denies that machines can think, one in effect assigns them to some lower level of achievement than that attained by human beings. But one might well contend that machines can't think, for they do much better than that. We could forever deny that a machine could think through a mathematical problem, and still claim that in many respects the achievement of machines was on a higher level than that attained by thinking beings, since machines can almost instantaneously and infallibly produce accurate and sometimes original answers to many complex and difficult mathematical problems with which they are presented. They do not need to "think out" the answers. In the end the steam drill outlasted John Henry as a digger of railway tunnels, but that didn't prove the machine had muscles; it proved that muscles were not needed for digging railway tunnels.

CHAPTER THREE

Robots, Consciousness, and Programmed Behavior[1]

I

But in spite of the foregoing, might we after all be a kind of robot? or might certain sorts of robots after all be a kind of us? Could a robot which simply did or had what it was supposedly programmed to do or have, be a robot which really felt and was conscious, had thoughts and intentions, deliberated and decided?

In *L'homme machine* La Mettrie had written:

> We are in the position of a watch that should say (a writer of fables would make the watch a hero in a silly tale): "I was never made by that fool of a workman, I who divide time, who mark so exactly the course of the sun, who repeat aloud the hours which I mark. No! that is impossible."[2]

So I shall apply the label "defenders of La Mettrie's watch" to those who claim it could never be shown that our own mental life and behavior is watch-like or programmed and that if a subject's mental life or behavior is watch-like or programmed, then it cannot be like ours. From Descartes to Cybernetics one can tick off defenders of La Mettrie's watch. In his *Principles of Philosophy* Descartes wrote:

[1] Some of the points developed here were suggested in my "Interview with a Robot," *Analysis,* 23 (1963), pp. 136–42.

[2] *Ibid.,* p. 145.

> We do not praise automata for precisely carrying out all the movements for which they were designed, since they carry them out by necessity; we rather praise the maker for fashioning such precise machines, because he fashioned them not by necessity but freely.[3]

And about three hundred years later Paul Ziff, in his provocative little article "The Feelings of Robots," wrote:

> MacKay has pointed out that any test for mental or any other attributes to be satisfied by the observable activity of a human being can be passed by automata. And so one is invited to say what would be wrong with a robot's performance. Nothing need be wrong with either the actor's or a robot's performance. What is wrong is that they are performances.[4]

He then goes on to suggest reasons (similar to Descartes') why "no robot could sensibly be said to feel anything":

> Because there are not psychological truths about robots but only about the human makers of robots. Because the way a robot acts (in a specified context) depends primarily on how we programmed it to act. Because a robot could be programmed to act like a tired man when it lifted a feather and not when it lifted a ton.

So too Jonathan Cohen in his article "Can there be Artificial Minds?"[5] marshalls a variety of arguments to support his contention that if a husband came to know and predict the future behavior of his wife in the way in which we can come to control and predict (in principle, at least) the future "behavior" of a computer (with or without randomising elements):

[3] In Descartes, *Philosophical Writings*, ed. and trans. Elizabeth Anscombe and Peter Thomas Geach (Thomas Nelson and Sons Ltd., 1954), p. 188.

[4] *Analysis*, 19 (1959), pp. 67–68.

[5] In *Analysis*, 16 (1965).

> . . . we should then say, I think, that his wife had no mind of her own, that her husband had already explicitly or implicitly done her thinking for her, and that the notion of programming an ideal mind, was therefore self-contradictory.

This conclusion will be of interest to us later. First, however, we shall be concerned with the jump Cohen makes from it to the following claims:

> What I have been trying to point out is an opposition between the still familiar concept of mentality and the concept of total subservience to known or knowable rules. The metaphor of "artificial minds" clearly destroys this opposition by ascribing mentality to programmed artifacts. That is why such a metaphor is–outside the laboratory–not merely eccentric and novel, but also undesirable. If we blur our concept of mind in this respect, then, however completely human beings were regimented in thought and action by their government's, party's, or church's ideological programme, it would be false to say that they behaved as if they had no minds of their own. They too could still be said to think for themselves. And that would be a shift in the notion of having a mind of one's own and thinking for oneself of which George Orwell's Newspeak could have been proud.

He goes on to write:

> But I do not see how our cyberneticist's unfortunate wife could ever recover a mind of her own. Her husband's difficulty–suppose he wanted such a recovery–would be logical, not technical. Whatever randomising elements he introduced into her brain, or however much he asked other cyberneticians to programme her and not to tell him what they had done, neighbours who were "in the know" would still nod their heads at one another as she walked past and murmur, "Poor girl, she can never again think for herself or have any emotions that are really her own."

Such arguments and attitudes have assumed a variety of forms and phrasings in recent years. But it will not matter if we neglect some of these nuances. For I intend to develop distinctions which will show that it would *never* follow from the fact of a subject being programmed that the subject failed to have thoughts, feelings, or intentions, perform this or that, etc. Insofar as (some) robots can be shown to lack minds it will be for different reasons.

I am more interested in *why* the defense of La Mettrie's watch fails than in *that* it fails. *Why* it fails has a bearing on current issues in the philosophy of mind and philosophical psychology, namely the relationship between consciousness and behavior and the explanatory limits of computer simulation of cognitive processes.

In Chapter Two I argued against what I took to be the most influential bad argument for saying that machines (or robots) could feel, think, or do such-and-such. In the present chapter I shall argue against what I take to be the most influential[6] bad argument for saying that machines (or robots) could *not* feel, think, and so on.

[6] Note Denis Thompson's recent article "Can a Machine be Conscious," where he refers to this argument as "probably the most common argument in the machine debate," *British Journal for the Philosophy of Science,* 16 (1965), p. 36. And compare as well Michael Scriven in his "The Compleat Robot: A Prolegomena to Androidology" where he begins Section 5 with a "quote" from the air: "Machines only do what we tell them to do, they are incapable of genuinely original thought." Scriven's objections to this view mark an about face from the stance he took in "The Mechanical Concept of Mind," in *Minds and Machines,* ed. Alan Ross Anderson, a fact often ignored by recent critics of that paper. What I say in this chapter is, at least in conclusions, much in accord with Scriven's later views. Also Turing, Putnam, Minsky, Armer, *et. al.* have at least sketched some dissatisfactions with the defense of La Mettrie's watch, and some of the things I say may be viewed as an elaboration of why such dissatisfaction has been justified.

II

First consider a typically whimsical philosophical preamble to a discussion of the mentality of robots. Suppose a skull is cracked open and some micro-wires etc. fall out. We become suspicious. We inspect closely the subject we took to be the owner of that skull. Very interesting. Finally we exclaim: "A robot . . . a programmed robot!" So we find *it* out. Whatever that subject had done in the past, whatever we had thought about it, "his" mental life however we knew it is now unveiled as no more than the exhibited capacities of a programmed robot. Things are seen in a new light, and some would say it is now clear that at best the robot only "behaved" and was "conscious."

So suppose we come to know that at one time so-and-so manufactured this robot, made this teeny mechanical brain which from the outset was endowed with a highly self-corrective micro-program (self-corrective in the way that certain programs for digital computers are, so that the machine may alter its routines to avoid previous blunders in a chess game) and "clothed and masked" it in such a way that it was "virtually indistinguishable from men in practically all respects: in appearance, movement, in the utterances it uttered, and so forth."[7] Now if we came to know such things, we might indeed with good reason think it necessary to revise some of our past attitudes toward and judgments about the subject who is now revealed as a robot. We would now suspect that certain descriptions and ascriptions which we had formerly thought appropriate of and to that subject were no longer appropriate. But exactly which ones, and why?

[7] Paul Ziff, *op. cit.*

Give or take a few details the preamble flickers out and a set of philosophical questions are posed. The central problem usually turns out to be the one of trying to determine which psychological predicates would no longer be applicable to the robot and which ones would remain essentially "intact." Let me call this *the repredication problem* for the unmasked robot. There is obviously no obvious solution to the repredication problem for robots given only the foregoing description of the unmasking. For this is a highly flexible Halloween surprise which could be satisfied by many different kinds of robots. Rather like cars and rather unlike people, robots come in a variety of makes and models. What we could say truly about a conventional shift may not extend to an automatic.

Questions which pop up at this point are "Is the programmed robot like a man under hypnosis? Or like a cunning dissembler? Or more like an actor with a script?" To assess these (often proffered) comparisons may be instructive. We find, for example, that the repredication that should occur in connection with the robot is very unlike the repredication which would be forced upon us were we to discover a subject to have done what he did as a result of being hypnotized (as, for example, where a man is hypnotized to insult someone who upon finding this out no longer feels insulted). Nor would it be like the case of the typical someone upon whom light breaks thus: "Look here, now, you've been deceived all along. She's only been pretending. She doesn't really love you." For to find out about the program in the robot's "life" (or life) is presumably, to find out something new about *all* the robot's "life" (or life). But what would it be to find that someone had been hypnotized for the whole of his life, or had pretended for the whole of his life? What would it be for anyone to live an *entire* life under hypnosis, or an *entire* life of pretense, or have someone's *entire* life shown to be performance? Compare: ". . . eight, nine, ten," snap: "You're no longer Descartes. You're just your old self." But *what* or *which* old self? Nor could one pretend to be Descartes all of his life. For when would the pretense begin? In the womb? Pretense, hypnosis, performance, though different from each other, are alike in this, that they all demand settings. They demand an environment in which they can occur and with which therefore they can be contrasted–a backdrop of non-pretense, non-performance. They must *enter into* the subject's life at a certain point, at a certain time. They cannot constitute the subject's mode of life at all points, at all times. The case of the robot and his program does not demand this to the same extent, which is one reason why Ziff is wrong (above) in his likening a robot acting according to a program to an actor's performance. Unlike the actor with a script, the robot we have imagined can easily be seen as always having been a programmed robot, which insofar as its actions are concerned has performed them in accordance with that program.

On the other hand it should be mentioned that only subjects with certain sorts of conscious capacities to begin with could be hypnotized, could be dissemblers, or performers with scripts. In other words, consciousness is presupposed by abilities to act in accord with certain kinds of directives. It is not a byproduct of such directions. What this reveals in connection with robots is that it would be wrong to suggest that a robot could be made conscious as a result of a foxy program. There is, to begin with, a subject, the robot, which has a program. There is not, to end with, a subject, the program, which is the robot. The importance of this morphology of mechanism is that what a program can be made to make a robot do will in part depend upon facts about the robot's nature, its hardware, its input potential, which are independent of its program. To assess *in toto* the mental machinations of a robot we need to consider not only how the robot is programmed, we need also to consider how and in what the program is roboted.

If the robot so loosely (but typically) characterized at the outset was neither conscious nor unconscious but non-conscious, then indeed we would have been deceived in any Buber-like "I-thou" relationship we had thought existed between us and that robot. Here the appropriate repredication would be more analogous to that which would occur were we to discover a ventriloquist had been deceiving us with his dummy. For example: "Charlie McCarthy seemed to be my most sympathetic friend until I found out he was made of wood. Fantastic. I felt and thought he had thoughts and feelings."

Compare: Joe intends to propose marriage to Martha in five minutes, and Martha knows this. A hypnotist swoops down on Joe and hypnotizes him to propose to Martha in five minutes. One marriage later Martha finds out about this. She should not cry out "Deceived!" She should simply smile at the coincidence.

A machine parallel to the latter case would be where a particular operation governed by a main routine was also governed by a subroutine. The subroutine instruction might be viewed as redundant as long as the main routine was working in an orderly fashion. But this would not mean that the subroutine instruction was *absolutely* redundant. For example, certain failures might occur in the execution of a main routine, so that the "spare tire" subroutine would carry out the instructions that the main routine failed to carry out. There is nothing inconsistent in conceiving of a stretch of behavior being guaranteed simultaneously by two different sets of directions.

Or let us suppose that a robot is designed and programmed to walk in a certain way. Now does the robot, could the robot, really walk or not? Contrast a robot such as the one imagined by Ziff, or Cohen's case of the programmed wife, with a wooden doll that trundles down an inclined plane. At best, such a doll does what resembles walking. Although we say, "Look at it walk," we are not (nor need be) careful in what we say–at least not philosophically careful. And by "philosophically careful" I mean only that we do not go on to say other things which we should say if harassed into preciseness. Here, however, I am not so concerned with whether it is blanketly inappropriate to describe the doll as one that walks, but whether, if we should so describe it, we should qualify and shield ourselves against certain misconstruals of our admission. If careful, we might rather point out that the walking toy only "walks" (not walks). Why not walks? Well, where else does it walk except down this board and others like it?

The trundling doll is hardly akin to Sisyphus and his eternal compulsory movements up and down and up that hill. It is not simply that in the case of the doll it is made to do what it does in the way of walking; it is also the case that the sort of walking it can be made to do is just about the only sort of movement it can be made to make. So even in this case it is not that we *make* it "walk," which counts automatically and decisively against contending that it *really* walks, or asserting that "of course it walks" without qualifications and guards. Rather, it is the sort of subject which we *make* "walk" which leads us to suggest "it doesn't really walk." This case, however, is more interesting than the ventriloquist and the dummy. For the doll really does do something–moves down the board. And the person playing with it does not get credit for its movements in the way in which the ventriloquist does get credit for what the dummy says. A person who causes a wooden doll to move down a board does not participate at every moment in the doll's movement down the board in the manner that a ventriloquist participates at every moment in the "conversation" of his dummy. The more detailed a subject's control of a second subject is, the more likely it is that we will credit the former with the doings that ensue as a result of his (or its) control over the latter. The more credit we give the controller, the less credit we may be likely to bestow upon the controllee for those same doings, given that we regard those doings as perforce the responsibility of one and not more than one agent. But where collective or multi-agent doings or actions are involved, the case is more complicated: each participant in the act may receive some credit for it, and no one participant need receive all the credit. For example, Allen Newell, J. C. Shaw, and H. A. Simon felt constrained (quite rightly, I believe) to list LT (initials for a program called the "Logic Theorist") as one of the authors of "Note: Improvement in the Proof of a Theorem in the Elementary Propositional Calculus" (C.I.P. Working Paper 8, January 13, 1958). Although Newell, Shaw, and Simon helped devise LT, they did not anticipate all the results that ensued from its use. Furthermore, control of a subject's doings need not be a specific movement-by-movement control. It may be more general, such as imparting to the subject a set of strategies and maneuvers that can be carried out in alternative ways. This also complicates the ascription of credit. Although I shall not develop these nuances, I mention them in order to cast suspicion on some assumptions that often ride piggyback on the defense of La Mettrie's watch: (1) that either the programmer gets all the credit for the doings of his robot *or* the robot gets all the credit; and (2) to control a subject through programming necessitates control over that subject's every movement every inch of the way. Neither of these assumptions are true. Just to find out that a subject is programmed, controlled, or directed in what he, she, or it does will seldom be sufficiently informative to permit us to decide on the stock of description(s) required to characterize correctly its mental life or lack of it.

In brief, if the robot had certain basic capacities to begin with, then the case of discovering him (her? it?) to be a robot might well not be a case where it would be appropriate to jump back and exclaim anything at all. If finding out about the robot is like finding out that a dog is not a dog but a wolf we may well feel that it would not be polite to exclaim (rudely) in front of the robot. Just as if we desist, for moral reasons, from beating dogs, we may desist, for the same (good) reasons, from beating wolves.

III

But here let me anticipate disappointments with what I have said so far. The strategy underlying most discussions of minds and machines is designed to uncover whatever conditions must obtain before it would be appropriate to ascribe feelings, thoughts, and so on to robots. This is done with an eye to tampering with our all too unclear perspectives with respect to human minds and their underlying makeup. But to be told that we should not withdraw predications of feelings, thoughts, and so on to robots *if* the robots had certain conscious capacities to begin with, then the case of discovering him (her? it?) to be nating. Yet I think it is—particularly in the context of recent philosophical discussions of mind and machines and computer simulation of cognitive processes.

Once the distinction is made between a robot's basic capacities, its input potential, on the one hand, and its program which may utilize these capacities on the other, the following claims seem in order:

If a subject is such that certain kinds of directives, commands, or controls (i.e. programs), could not be imposed upon him (her? it?) then, of course, the subject will not be able to perform various mental acts, infer this from that, prove theorems, write couplets, etc. But the reason will not be the programmed nature of the subject's mental life (whatever it amounts to), but will be the nature of the subject who (which) simply lacks certain sorts of programming in the first place.

On the other hand, if the subject is a certain sort of being to begin with such that certain kinds of directives, commands, and controls are possible, then the simple fact that he (she? it?) was programmed to do this or that will not be sufficient to establish that the subject did not really do this or that.

In the case of programmed *behavior* and those mental aspects closely associated with it the defense of La Mettrie's watch fails because with few exceptions there is no (clear or necessary) opposition between behavior and being programmed. Indeed, behavior seems particularly amenable to being construed in terms of well-defined tasks, and because of this it is compatible with being programmed. In their *Plans and the Structure of Behavior,* for example, Miller, Galanter, and Pribram write:

> Any complete description of behavior should be adequate to serve as a set of instructions, that is, it should have the characteristics of a plan that could guide the action described. When we speak of a Plan in these pages, however, the term will refer to a *hierarchy* of instructions, and the capitalization will indicate that this special interpretation is intended. *A Plan is any hierarchical process in the organism that can control the order in which a sequence of operations is to be performed.*
>
> A plan is, for an organism, essentially the same as a program for a computer, especially if the program has the sort of hierarchical character described above.[8]

Much current research into computer simulation of problem-solving rests on this intuition. Behavior insofar

[8] George A. Miller, Eugene Galanter, and Karl H. Pribram, *Plans and the Structure of Behavior* (New York, 1960), p. 16.

as it can be construed in terms of well-defined tasks harbors an inbuilt compatibility with being programmed (call it "program-receptiveness"). And if a subject fails to possess the basic capacities necessary to the construction of routines for proving theorems, writing couplets, etc., *then* the nature (or basic capacities or input potentials) of the subject determines that the subject does not do these things and pre-empts and precludes the possibility of the subject failing in these respects because he (it) was merely programmed to effect them.

Most simply put, and contrary to the defense of La Mettrie's watch, there is *no* type of mental predicate which fails to apply to a subject *just because* it was programmable in certain ways. Although various quasi-Rylean, behavioristically oriented, problem-solving psychological predicates may be shown *not* to apply to a given subject, this will be because the subject lacks the capacity for being programmed in certain ways. It will not be because the subject *was* programmed in certain ways.

And when we turn to those aspects of mentality which are quite different from solving problems, playing checkers, or writing couplets–such things as having pains, emotions, after-images, etc. . . . the defense of La Mettrie's watch may be seen to be wholly irrelevant. For the having of pains, emotions, after-images, etc., are all examples of non-problem-solving non-behavior. They are not potentially well-defined tasks which hence may be programmable, for they are not tasks at all. Consequently they could never be shown to be absent from an unmasked robot *because* the robot was merely programmed to do this or that. For they are not the sorts of things which a robot could do. They may be had, but not done.

Whether or not a robot has a capacity for pain, emotion, or after-images, will of course greatly effect what the robot could *then* be programmed to do. The problem of determining the nature of the subject is not primarily a problem of deciding how to program it. It is a problem of knowing and deciding how to construct the subject so that it *can* be programmed in certain ways.

Thus we see that the robot without a program need not be a robot without a mind. For it could be a robot with half a mind—a robot with certain capacities for sentience, with certain input potentials, etc. It would by nature be a non-sapient robot . . . submoronic when compared to others of the same model which had also been equipped by benevolent manufacturers with programs which use their basic capacities.

On the basis of the preceding remarks, then, it may be useful to make a preliminary sorting of psychological features into two kinds: those which it at least makes sense to attempt to simulate with a machine or robot through programming (in the sense of devising routines), and those which it does not. The first I shall call *program-receptive* features of mentality. These include most kinds of problem-solving: game-playing, theorem-proving, etc. . . . i.e., rule governed activities. The second set of features I shall call *program-resistant.* These include such things as having pains, feelings, emotions, etc. This distinction will be the focus of Chapter Five. But for our immediate purposes it is sufficient to point out that it would be a methodological howler to attempt to simulate pains, feelings, and emotions with a machine simply by expanding current programming techniques in the sense of defining new routines. It may of course prove possible to produce machine analogues of these features through a development in hardware. (There are undoubtedly mental aspects in between those which are clearly *program-receptive* and those which are *program-resistant* —e.g., thinking. But I shall not go into that here.) Thus some psychological predicates may be viewed primarily as software predicates; some primarily as hardware predicates.

It is unfortunate that this rather elementary but crucial distinction is very often overlooked by both philosophers and those working on artificial intelligence and computer simulation of cognitive processes. This has resulted in philosophers coupling the mental limitation of machines (period) with the limitations as to what a machine could be programmed to do. And, as we shall see, it has resulted in some computer simulation theorists trying to simulate *program-resistant* features of mentality using simulation techniques which are strictly relevant only to the simulation of *program-receptive* features.

To show that such a machine was only programmed to do such-and-such could not possibly show the absence of consciousness in that machine. Conversely, however, one could not make such a machine conscious simply by programming it in a certain way. Instead the link between consciousness in machines and what machines can be programmed to do is this: if a machine could be programmed in certain ways, this might demand consciousness in the machine; and if it were not possible to program the machine in certain ways, this might suggest that the machine lacked capacities which we can loosely refer to as its consciousness.

IV

The above should help us to see why Ziff's remarks about how a robot's actions in a particular context depend on how it was programmed to act have little bearing on the question "Could robots have feelings?" Looked at from one angle it seems that Ziff's imagined robots in the last analysis are not capable of feeling anything. But if so, the only reason they are not, is that they are specified in a vague way at the outset of Ziff's article and end up, as it were, "retroanalytically defined" as robots which are without feelings. That is to say, at the end of his analysis we are told what his robots, about which so many interesting questions were raised at the outset, were really like. But what I have tried to show is that other "retroanalytic definitions" are compatible with the robots so flexibly characterized at the beginning of Ziff's article.[9] And furthermore and mainly what I wish to object to is the suggestion that Ziff's robots, however interpreted, cannot feel *because* they only do what they are programmed to do, or *because* they could be programmed to respond in all sorts of strange ways to various stimuli. Ziff's programmed robots cannot feel simply because of the sorts of robots Ziff decided to say that they are; not unsimply because of the sorts of programs they have. From another angle, however, Ziff's view becomes virtually impossible to understand. This is so if one attributes to him the claim attributed to him by Putnam where he (Putnam) says: "Ziff has informed me that by a 'robot' he did not have in mind a 'learning machine' of the kind envisaged by Smart, and he would agree that the considerations brought forward in his paper would not necessarily apply to such a machine (if it can properly be classed as a 'machine' at all)".[10] Given this interpretation it is difficult to understand how Ziff's robots could be "virtually indistinguishable from men in practically all respects: in appearance, in movement, in the utterances they utter, and so forth." Ziff wishes to have it both ways if he does not want his robots to be as sophisticated as "learning machines,"

[9] Compare this criticism of Ziff with a criticism of Ayer's "The Concept of a Person" made by Raziel Abelson in his "Persons, P-Predicates, and Robots," *American Philosophical Quarterly* 3 (1966), p. 308.

[10] Hilary Putnam, "Minds and Machines," in *Dimensions of Mind*, ed. Sidney Hook, p. 176.

and yet seem as sophisticated as men. Even to *seem* that sophisticated in mental capacities strongly suggests subjects psychologically less vacant than Ziff supposes his robots to be.

Similarly, Cohen's remarks on artificial minds are mistaken insofar as he claims an opposition between *mentality* and the concept of total subservience to known or knowable rules. All that Cohen's case of the totally programmed wife shows is that someone's mind could be controlled in a certain way: it does not show that a wholly programmed mind is no mind at all. To show that someone does not think for himself, is not in the least to show that someone does not really think, and in the same sense(s) as the present sense(s) of "think," whatever that (or they) may come to.

Arguments to the effect that a robot which only did what it was programmed to do could not be a robot which *really* was conscious, had feelings, thoughts and intentions, or which *really* deliberated and decided, fail because of the foregoing. But here let me anticipate some objections to my objections. One reason why the notion of "being programmed" has come to seem opposed to the notion of "being conscious" is because most of the currently existing programmed devices (computers, *et al.*) have appeared—at least to some philosophers—obviously non-conscious, so in fairness to part of the "spirit" of some of the articles being criticized, I should add that there is some textual justification for pointing out that the authors of the articles cited are not always arguing simply from the fact that a subject is programmed to the conclusion that the subject is therefore non-conscious. They are also, to some extent, presupposing that there are other relevant facts which one discovers when one discovers that the subject is programmed which count against saying that the subject could be conscious. Nonetheless, (1) throughout the literature these presuppositions are never clearly set out, defended, or developed; (2) no distinction is drawn between the very different sorts of subjects which could be programmed; (3) there is no reason for supposing that for every subject which is programmed one will be able to uncover independent facts that would tend to establish that the subject was not conscious, capable of thought, creativeness, and the like. And (4) in some cases, given that the subject could be programmed to do certain things, what we would expect to find would be just the opposite: namely that the subject *was* conscious, capable of thought, intentional behavior, and so on. (Exactly what it would take to show that such expectations were satisfied is, of course, no easy matter to decide. The "other minds" problem could arise in connection with robots—even though we made them.)

V

The foregoing points may be summarized by drawing some parallels with the late J. L. Austin's article called "Pretending" where he wrote:

> The moral is, clearly, that to be not-pretending to be, and still more to be not only-pretending to be, is not by any means necessarily, still less *eo ipso,* to be really being. This is so even when the way in which we fail to be (only-) pretending is by indulging in excessively "realistic" behavior: but of course there are also numerous other kinds of case, some to be mentioned later, in which we might be taken to be pretending and so may be said to be not pretending, where the reasons for which we are said to be not pretending, where the reasons for which we are said to be (only-) pretending are totally different from this, and such that the notion that not-pretending really being could scarcely insinuate itself. We must not allow ourselves to

be too much obsessed by the opposition in which of course there is something, between pretending and really being: not one of the following formulae is actually correct:

(1) not really being ⊃ pretending
(2) pretending ⊃ not really being
(3) not pretending ⊃ really being
(4) really being ⊃ not pretending[11]

I am not here concerned with defending or recounting Austin's arguments, and nothing which I assert depends on the truth of Austin's claims. All I wish to do is exhibit a set of claims which differ from Austin's only in that they are phrased in terms of a dichotomy between programmed behavior and real behavior: From the simple fact that a subject S (robot or machine) is programmed to do *X* it cannot be shown that the doing of *X* was in some way unreal, artificial, phony, not genuine, and so on. If this is so, then arguments of the form "The robot didn't fall in love, it only did what it was programmed to do" fail.

Following Austin's scheme, I would assert that in the case of a robot (or any other subject for that matter), not one of the following formulae is correct:

(1) not really being ⊃ being programmed
(2) being programmed ⊃ not really being
(3) not being programmed ⊃ really being
(4) really being ⊃ not being programmed

To show that (3) is not correct we need only imagine the case of a robot that is not programmed to do what it does but, due to an accident in its circuitry, emits a string of phonemes which sound as if it is really asking questions, though it is not. To show that (2) in my scheme is not correct, we need only imagine the case where a robot is programmed to move (we will not even say "walk") across the room and does. A digital computer that calcu-

[11] *Philosophical Papers* (London, 1961), p. 205.

lated such-and-such because it was programmed to calculate such-and-such would also suffice to show the inadequacy of (2). To show that (4) is not correct we need only imagine again the robot moving across the room because it was programmed to move across the room. Similarly (1) is not correct because of the aforementioned possibility of the robot not really asking a question, though it appeared to because its wires got crossed.

VI

It may be argued that though it would be possible to program a robot to have or effect all sorts of marvelous (humanoid) experiences and accomplishments if it had certain basic capacities to begin with, it is an analytic truth that robots could never possess the requisite capacities for such experiences or accomplishments. But those who would argue that a subject could not be both a robot and have certain basic capacities or input potentials, may be in the same position (now) as James and others were in toward the end of the nineteenth century when it was vehemently maintained that a subject could not be both wholly mechanistic and exhibit self-adaptive behavior. This also had the flavor of an analytic truth before various negative feedback machines were developed and the notion of "self-adaptive behavior" received further clarification.[12]

If psychological predicates are really applicable only to persons, which I think is unlikely, then robots of certain sorts might be persons of certain sorts. They would be persons to which certain other hitherto non-person predicates applied—certain clusters of what might be

[12] Cf. my "Cybernetics," in *Encyclopedia of Philosophy,* ed. Paul Edwards (New York, 1967).

called B-(biographical) and E-(ecological) predicates such as "is manufactured," "was a 1966 model," "sleeps electrically," etc.

There is, however, no more reason to assume that robots would have to be persons in order to partake of psychological predicates than there is reason to assume that a dog would have to be a person to be able to show affection and obeisance. In other words, I believe that psychological predicates can become applicable to machines or robots, as I think some already have, without blurring the line between robots and persons. In short it seems possible that such a line need not be blurred *except with respect to the predicates in question.* That is, it may be possible to show that a machine could, for example, recognize various sorts of patterns without blurring the line between men and machines *except insofar as recognition is concerned.* If in such a case the line between men and machines need not be blurred except with respect to "recognition" (and its near synonyms) then it is hardly clear that the meaning of such words as "men," "persons," "robots," or "machines" will have altered. Both rolling stones and rolling men gather no moss, which hardly shows that men are borderline cases of stones. When Thomas Edison invented the light bulb he did not produce a borderline case of a white dwarf star. (Compare: "If it really throws light, it's some sort of star" with "If it really has feelings, it's some sort of living organism.")

In the case of living things and machines there was no global blurring of the line between men and machines when we began (appropriately) to refer to "self-adapting" machines. Instead a localized sharing occurred in connection with the feature of self-adaptation. That is precisely why cybernetics seemed to provide a non-trivial counter example to some of the claims made by vitalists.

So to build robots to which all sorts of psychological

predicates legitimately apply need not be comparable to building robots which were just like us. They would of course be rather like us insofar as they share those predicates. And to be like us with respect to sharing a given capacity or the exercise thereof does not necessitate that the robot be identical to us even in that respect. Compare: Both washing machines and human beings wash clothes, but they are not identical with respect to this capacity or the exercise thereof. Hence for a robot to have psychological predicates apply to it is not automatically, as many would assume, either the same as the robot being the same as, or even being almost the same as, a person.

To treat philosophical questions such as "Can machines think?" as primer problems in "conceptual analysis" underrates the possibility of "conceptual reanalysis" —or what takes place when discovery and invention (say, of new machines) *forces* us to reexamine and reassess what certain words ("machines" *et al.*) could apply to if not in the first at least in the second place.

When Wittgenstein wrote:

> But a machine surely cannot think!—is that an empirical statement? No. We only say of a human being and what is like one that it thinks. We also say it of dolls and no doubt of spirits too. Look at the word "to think" as a tool . . .[13]

he was mistaken. He was mistaken because he assumed one can make a conceptually pure non-empirical point (about the current use of "machine," "think," *et al.*) and settle the matter. No conceptual analysis based on current usage, dictionary entries, and so on can (now) settle the matter unless it can somehow anticipate any new discovery, invention, and so forth that might take

[13] Ludwig Wittgenstein, *Philosophical Investigations,* trans. G. E. M. Anscombe, Oxford, 1953, p. 113e.

place. That would be an advance in diachronic linguistics too trusty to be true.

To put it another way, it is not that strictly speaking we know Wittgenstein to be wrong in the above paragraph, it is simply that he could not have known that he was right. For there is no assurance (now) that the above quotation will not some day appear to us as the following paragraph now appears:

> But a machine surely cannot play chess!—is that an empirical statement? No. We only say of a human being and what is like one that it plays chess. We might imagine a doll or perhaps a spirit playing chess. Look at the phrase "plays chess" as a tool.

But now this remark would be false.

Furthermore, it is not even the case that in every instance where the concepts were blurred, interesting discoveries would be ruled out. For how the concepts (of men and machines) come to get blurred is also of central importance. If, for example, we found that we do not really do some things which we thought we did, and are therefore more like certain machines than we had hitherto realized, or if we found out that the way that we did things was more purely mechanized than hitherto imagined, then in both cases something very interesting might be found out. (Compare recent suggestive attempts by Newell, Shaw, and Simon to explain creative thinking in human beings in terms of certain kinds of rule-governed behavior.)[14]

A view more explicitly of the kind the immediately foregoing remarks contrast with is developed by A. R. Lacey in his article "Men and Robots" where he writes:

> Let me add again that I do not mean that no artifact will ever achieve feeling, but that such an artifact would not be

[14] "The Processes of Creative Thinking," RAND Corporation, P–1320, September 16, 1958.

a robot, in my sense, but a human, or on the human side of the fence.[15]

But it is not at all certain that in every case where artifacts had feelings they should for that reason be placed "on the human side of the fence" (except trivially, for example, with respect to feelings). And I disagree with him where, after despairing of finding the differences between men and robots in terms of what they can do, he writes:

> This does not mean that we could not create an artifact which turned out to be conscious; but such an artifact would be a human (or some such thing) and not a robot in our present sense.[16]

His qualifying phrase "in our present sense" does not barricade Lacey's claims against the linguistic possibility that "conscious" could acquire new uses (without altering its meaning)[17] in much the same manner that "problem solving" and "self-adaptive" seem to have done. It remains possible in the face of such diachronic surprises to trivialize a thesis like Lacey's by, say, simply not identifying the present sense of "robot" with the then current senses of "robot." One can always choose to use a word in an idiosyncratic manner, but nothing is gained thereby. (There is no such thing as an idiosyncratic *sense* of a word.) And obviously Lacey is not doing that. So if he does wish to identify his construal of the meaning of the

[15] *Philosophical Quarterly*, 10 (1960), pp. 61–72.

[16] *Ibid.*

[17] Cf. R. Puccetti's interesting article "On Thinking Machines and Feeling Machines," *British Journal for the Philosophy of Science*, 18 (1967), where a parallel use is made of Putnam's important distinction between (1) a word acquiring a new *use* in the language because of the (core) meaning it has (had) and (2) a word being (arbitrarily) *given* a new use (or meaning). And see my "Cybernetics and Mind-Body Problems."

word "robot" with *the* meaning of the word "robot," his claims based in part on this identification will be vulnerable.

There need not be any "the difference" between men and robots. Different differences are possible and must be discussed in any attempt to stake out what a robot could be and do and how what it could be or do would serve to distinguish men from robots. Some of these differences would no doubt be marked out by the use of psychological predicates and some would be marked out by the use of what I have loosely called B- and E-predicates. Some by others.

Many philosophers including many defenders of La Mettrie's watch have looked for an almost magical covering word or predicate (or small set of predicates) such that if it or they (always) applied or failed to apply to a certain subject we would then *be safe* (always) in saying that the subject was definitely (or definitely not) a human, or a robot, or a machine. And so they persist in perpetrating such claims as "If a subject could think (create, feel, perform, purposively, and so on) then it would not be a robot (or machine, etc.)" and have failed to note that no one feature (or small set of features) such as "feels" or "is made in a factory" can be regarded as essential to something being or not being a robot or human being or machine. Perhaps a programmed conscious robot could perform as well as a human being, and do this by, for example, thinking out the answer, and still not thereby be a kind of us. Many other relevant differences might remain: biographical, ecological, and so on.

VII

Think of the H's in the boxes below as representing predicates that are (clearly) applicable to human beings.

These may be ecological or biographical such as "born of woman" and so on. Think of the R's as representing (clearly) robotic or machine predicates, such as "designed and built by IBM, programmed by Newell, Shaw, and Simon" and so on. And think of the P's as representing various so-called person predicates or, less controversially, psychological predicates:

	1	2	3	4	5	6	7	8	9
1	H	H	H	P	P	P	R	R	R
2	H	H	H	P	P	P	R	R	R
3	H	H	H	P	P	P	R	R	R
4	H	H	H	P	P	P	R	R	R
5	H	H	H	P	P	P	R	R	R

#1

(The dotted lines are meant to indicate the actual indefiniteness in the number of each kind. We shall, however, ignore this in the ensuing discussion. We shall also ignore the possibility of discovering that some H's are really R's or that some P's are really R's—for example, that thinking is identical with executing a programmed routine. I am also assuming a principle for distributing H's, P's, and R's in their respective areas. For example, sentience predicates might be located in one part of the P-area, sapience predicates in another.)

The question then arises as to which combinations of H's, P's, and R's are possible and which are not. It seems, for example, that no subject would be characterized by all H's and no P's unless the subject were deceased. On the other hand is there any argument to show that necessarily no P's would be applicable to any subject largely characterized by R's? If problem-solving predicates are in some sense psychological predicates, then this single fact would, it seems, be sufficient to undermine an affirmative answer to this question.

Some subjects, according to my account, might be describable by the following predicate distribution "shape":

	1	2	3	4	5	6	7	8	9
1	H	H	H	P	P	P			
2	H	H	H	P					
3	H	H	H	P	P				
4	H	H	H	P					
5	H								

#2

And some might for all we know be describable by:

	1	2	3	4	5	6	7	8	9
1				P	P	P	R	R	R
2						P	R	R	R
3					P	P	R	R	R
4						P	R	R	R
5									R

#3

whereas other subjects might be described in a more "hybrid" way by:

	1	2	3	4	5	6	7	8	9
1	H	H	H	P	P	P	R	R	R
2			H	P	P	P	R		
3				P	P	P			
4				P	P	P			
5					P				

#4

A subject described by shape #2 is meant to correspond to our intuitive notion of a human being to whom a fairly wide range of psychological predicates apply. (A moron, by contrast, could be depicted by a large number of H's and fewer P's.) A subject described by shape #3 is meant to correspond to our intuitive (and, I believe, perfectly consistent and coherent) notion of a robot with a noticeable sprinkling of mental capacities. And diagram #4 is meant to correspond to our intuitive notion of a borderline case of humans or robots or what I would call "persots" or "robsons."

The positions I have wished to reject (such as Lacey's and others') have in common their flirtation with the view that any subject to whom a P applied is necessarily a subject to whom primarily H's also apply, and any subject to whom primarily R's apply is necessarily a subject to whom no P's apply. Combinations such as

	1	2	3	4	5	6	7	8	9
1	H	H	H	P	P	P	R	R	R
2				P	P	P	R	R	R
3							R	R	R
4									R
5									R

#5

and an impressive myriad of others would be (analytically) ruled out. I have yet to see any version of this view whose rigor of formulation is even remotely commensurate with its alleged power to consign to the bin of the impossible any "shape" that included a P and that was not clearly dominated by H's.

VIII

True borderline-case decisions may be as rare in philosophy as true tied-finishes are in the Kentucky Derby. So far, at least, the question of whether an actual subject is a humanoid machine or a mechanistic human has not arisen. Instead the question has arisen as to whether certain predicates ("recognizes patterns," "plays chess," "writes poetry," etc.) hitherto used only in connection with human beings should apply to what are most certainly machines; or whether certain predicates hitherto used only in connection with machines (e.g., "is programmed") also apply to what are most certainly human beings.

In his challenging article "Robots: Machines or Artificially Created Life,"[18] Hilary Putnam seems to me to mischaracterize the central issues involved in "minds and machines" controversies insofar as he assimilates the question of clarity "with respect to the 'central area' of talk about feelings, thoughts, consciousness, life, etc." to the question of clarity "with respect to the 'borderline-case' of robots (i.e., as machines or artificial life)." If and when we draw open the closed curtain of technological ignorance and find persots or robsons inhabiting the parlors of persons, we will probably (though not necessarily)[19] know what to say about them. If problems persist which are to be solved by making a linguistic decision, they will be trivial problems: for in a clear case of a borderline case the only decision involved in selecting descriptions is a poetic one (e.g., shall we call them "persots" or "robsons"? I myself prefer "persots"). To think otherwise is to mistake neology for ontology. A tangerine is a tangerine is a tangerine.

The non-trivial, non-science-fictional problems of philosophy now are rather such problems as: Are the information processes underlying the problem-solving capacities of, for example, a (clearly non-human) list-processing computer analogous to human ones? Do so-called pattern-recognition programs actually enable machines to *recognize* patterns? What sorts of evaluation procedures can be developed which will assist us in deciding which of any two competing simulations are the most useful in speculating about human mentality. Which machine models of human mentality are subject to plausible neurophysiological interpretation, which ones are not? What relationships obtain among mentalistic phenomena which are receptive to being simulated by current computer programming techniques, and those which are not? Which psychological predicates refer to basic capacities, and which apply to

[18]*Journal of Philosophy*, 61 (1964), pp. 668-91.

[19]Cf. my "Cybernetics and Mind-Body Problems," *Inquiry*, 12 (1969), pp. 406-19.

behavior "generated" by those capacities? Which are a mixture? Short of making a mind, how can we model aspects of the mind?

In other words, the problem we presently face is the one of trying to understand how our own understanding of a subject corresponding to

	1	2	3	4	5	6	7	8	9
1						P	R	R	R
2						P	R	R	R
3							R	R	R
4							R	R	R
5							R	R	R

#6

might increase our understanding of subjects corresponding (in significant degree) to:

	1	2	3	4	5	6	7	8	9
1	H	H	H	P	P	P	?	?	?
2	H	H	H	P	P	P	?	?	?
3	H	H	H	P	P	P	?	?	?
4	H	H	H	P	P	P	?	?	?
5	H	H	H	P	P	P	?	?	?

#7

namely, ourselves.

It is this cluster of topics to which we now turn.

CHAPTER FOUR

Philosophy and Computer Simulation

I

One less than sunny assessment of the theoretical advantage to be derived from the employment of mechanistic models in a study of the mental is to be found in Kenneth Sayre's *Recognition,* where he writes concerning machine analogs of human recognition capacities:

> We simply do not understand what recognition is. And if we do not understand the behavior we are trying to simulate, we cannot reasonably hold high hopes of being successful in our attempts to simulate it.[1]

But *contra* Sayre,[2] I find it unsatisfactory to maintain that "if we do not understand the behavior we are trying to simulate, we cannot reasonably hold high hopes of being successful in our attempts to simulate it." (Here for the sake of exposition let us assume as Sayre does that the goals of most pattern-recognition research have been to *simulate* human pattern-recognition capabilities.)

Now certainly the main motivation for this has been to further our understanding of the behavior and/or psychological process to be simulated. To require detailed understanding of the phenomenon to be simulated prior

[1] *Recognition: A Study in the Philosophy of Artificial Intelligence* (Notre Dame, Ind., 1965), p. xii.

[2] I believe that Sayre no longer holds this position. Nevertheless, it is a naturally tempting one that ought to be considered.

to devising the simulation is to guarantee that the simulation be theoretically irrelevant. Its utility in psychological explanation will reduce to that of a sort of visual aid—a method for displaying what is already known.

The point is that the theoretical efficacy of computer simulation depends on our having some knowledge of the phenomenon to be simulated, *but not too much.* Preliminary clarification of the nature of the phenomenon to be simulated is no doubt required. But this more as a way of charting the course than of making the journey. Sayre seems to have assumed that only after we have clarified the concept of recognition will we understand how to program a machine with recognition capacities, whereas I want to suggest that by trying to program a machine with recognition capacities we will perhaps improve our understanding of the concept of recognition—and, of course, of recognition itself.

(A book such as Ryle's *Concept of Mind* seems to me best appreciated as an attempt to arrive at a preliminary clarification of mental phenomena. And it may be that the enormous difficulties in doing even this well have lead some philosophers to assimilate philosophical activity *per se* to the kinds of analytical tactics employed in that work.)

Sayre's (early) work cited above presents a somewhat curious mixture of attitudes in this regard. On the one hand there are remarks such as the last quoted. Yet on the other hand he openly uses Rylean type distinctions (between achievements, attainments, and processes) as a preface to and not a substitute for working on simulation programs. One drift of his argument against pattern-recognition studies is that if such researchers had read their Ryle, they would not have used interchangeably an "attainment" verb such as "to recognize" with verbs such as "to classify" or "to identify." The latter are said to range over behavior admitting of greater or lesser degrees of success, whereas recognition is said to be "evaluation neutral"–something not done more or less well. Clearly, unless one is able initially to specify the phenomena he wishes to simulate, it is highly unlikely that the desired simulation will be brought off. And obviously, conceptual analysis (what else?) can be an aid in carrying out the initial specification.

It would, of course, underrate the philosophical importance of computer science to suppose the conceptual insights (into "problem-solving," "self-adaptiveness," etc.) alluded to in our Introduction would have been gained without the technological illustrations. They might have been, but what is important is that they were not. Yet it would overrate the philosophical implications of those same types of illustrations to suppose they provide a kind of panacea to most previously intractable problems in philosophical psychology.

As we have seen, in their informal discussions of such questions as "Can machines think?" and "Could a robot have feelings?" philosophers have sought to clarify our hazy intuitive notions regarding the mental life. Simulation psychologists, by contrast, have focused on the details of actual programming and hoped thereby to grasp in detail the ways in which it is or is not possible to model mentality with current computers.

The two tactics are surely quite compatible, if not inseparable, and a formal marriage between philosophical analysis and computer simulation seems desirable.

But I certainly agree with Sayre to this extent: unless *some* initial clarification of the nature of the phenomenon to be simulated is attempted, the simulation is likely to lack direction and its evaluation will seem arbitrary. As Walter R. Reitman wrote:

> Although we are interested in natural phenomena, we can only work with what we have ways of thinking about. The

trouble arises when we try to discover just what our constructs simulate.[3]

One way to characterize the implications that computer science has for philosophy is to say that it provides for certain problems new and sharper ways of rephrasing them—namely, in terms of programming problems. To rephrase is not, of course, to solve. But it is sometimes a useful first step to a solution. We come to see more clearly the edges of our ignorance.

In what follows I shall try to illustrate how certain programming problems may be viewed as rephrasals of certain philosophical problems. I shall then make some critical comments on the craft of computer simulation as well on the comments of certain (philosophical) commentators who have for different reasons been critical of it.

But first some orienting remarks on pattern-recognition, computer simulation, and artificial intelligence. (Note: Most problems that computer simulation programs deal with, such as theorem-proving, chess-playing, etc., are problems in pattern-recognition of one sort or another. Sometimes in the literature, however, pattern-recognition is equated with letter- or character-recognition (in cursive script, say). It should be clear from the context exactly what sort of pattern-recognition problem is under discussion.)

II

Pattern-REcoGnition, Computer Simulation, and Artificial Intelligence

A discussion of PATTERN-ReCogNition is, of course, a discussion of Pattern-Recognition, which is, of course, a

[3] *Cognition and Thought* (New York, 1965), p. 6.

discussion of pattern-recognition, and hence we have exhibited the phenomenon many would like to explain.

Roughly, the aim of pattern-recognition research is to devise mechanical means whereby non-uniform instances of an auditory or visual pattern, the letter "M," say, would be distinguished as M's. As Newell and Simon have pointed out, however ("Computers in Psychology"),[4] it is possible to view a pattern-recognizer as virtually any system that makes different responses to different stimuli. But it will prove satisfactory for our purposes to confine our examples to those where a machine has been programmed to recognize a given character or characters of the alphabet.

To follow Newell and Simon, the possible stimuli might be combinations of ones and zeros in a large two-dimensional grid of ones and zeros:

Figure 1

1001011000
1010110100
1011101011
0101011101
1101101011

A particular character of the alphabet is then equated with particular subsets of ones and zeros. Thus the following stimuli might be M's:

Figure 2	*Figure 3*
11100111	10100101
11011011	10011001
11000011	10000001
11000011	10000001

[4] In *Handbook of Mathematical Psychology,* Vol. I, ed. R. Duncan Luce, Robert R. Bush, and Eugene Galanter (New York, 1963), pp. 361–428.

Consequently the goal for the learning system is to distinguish various letters of the alphabet which could appear on the grid in different positions, sizes, orientations, and styles. Cast in these terms, human recognition of patterns may be viewed as the ability to recognize given instances of the same character, each instance of which may differ widely from the others with respect to either position, size, orientation, and style.

As an exercise in AI, successful pattern-recognition (with respect to given characters of the alphabet) would consist in emulation of the results of exercising this ability. As an exercise in CS, a successful pattern-recognition program would simulate the *process* by which that ability was exercized.

It was pointed out in the Introduction that the aims of AI are to devise mechanical means for executing tasks that hitherto required the actions of intelligent agents. Hence there need be no attempt to simulate whatever intelligent process or procedure was used by human beings if it is possible to bring about the same results in other ways. Since the aims of AI are essentially practical, replication of human processes or procedures would be happily sacrificed for speed and efficiency. If there are ways of translating faster and more efficiently from Russian to English than those ways used by intelligent agents, so much the better. The goal is not to imitate the mind but to imitate what the mind can do. CS, on the other hand, is directed towards modeling the mind. The goal is to simulate the mental processes at work in human thought and action.

To follow Newell, Shaw, and Simon,[5] we can let a simulation of an alleged mental process be a computer program the flow chart of which is (generally) constructed on the basis of observations (verbal reports,

[5] "Computers in Psychology."

marks on paper, and so on) of intelligent behavior such as playing chess or checkers, proving theorems, and the like. The program, when run on the computer, can be instructed to print a trace of its move-by-move activity which may then be compared to a putative description of the process.

As we shall see, a large number of as yet unsolved problems arise in trying to assess the theoretical importance of this tactic. But if we suppose the program for the simulation is written in Information Processing Language-V (IPL-V)—a language that works with lists and lists of lists of symbols, then the general theoretical situation may be depicted as follows:

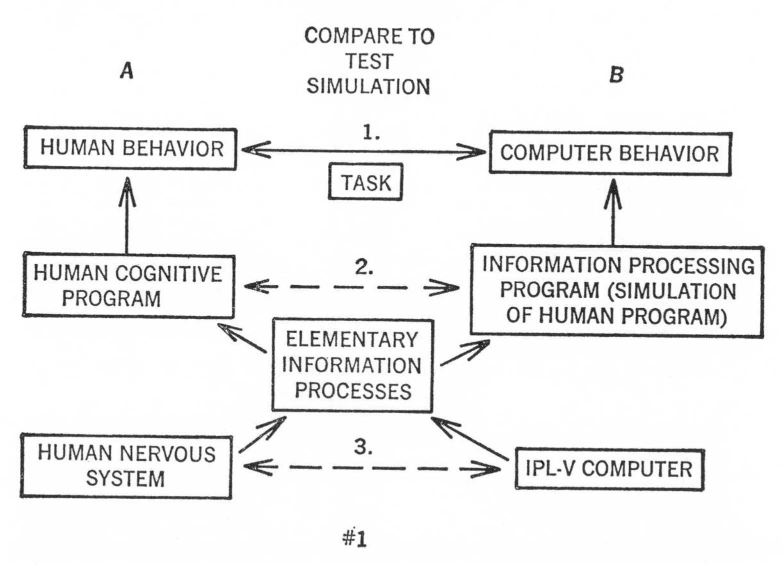

#1

where, at the moment, for reasons some of which will be discussed later, detailed comparison between the human and machine system, insofar as there are any, exist for the most part at the task level 1. Hence the solid-line arrow, which indicates that at least some comparisons can be made. The broken-line arrows 2 and 3 are meant to depict the general paucity of comparisons between items at that level in columns *A* and *B*. This lack is partly because it is not even known whether human beings can be accurately described as behaving in accord with programs or how these might be depicted, etc. It is hoped, however, that task-level comparisons can be refined, properly assessed, that deeper comparisons will eventually be forthcoming at other levels (2 and 3), and that suggestive analogs to the mechanisms and processes known to be sufficient for producing certain machine outputs will be discovered for human outputs. It is also worth asking whether various vertical relationships between items in column *B*, such as how a program relates to its machine's hardware, would be paralleled by the vertical relationships between items in column *A*. (For a further discussion of #1, p. 97, which with the exception of the broken-line arrows is due to Newell and Simon [*op. cit.*], see my "Cybernetics and Mind-Body Problems.")

What I wish first to focus on, however, are some problems in epistemology which have recently been paraphrased as problems in computer programming, for the extent of our failure to solve these problems (under either phrasing) probably represents the extent to which we fail to understand understanding. And this, obviously, has an important bearing on the extent to which we might mechanistically model the mental.

III

Pattern-Recognition, Implicit Recognition, Family Resemblances, Relevant Projection Capacities, and the Curse of Context.

I shall begin with another claim of Sayre's. He writes:

Yet we have argued that few would be able to say exactly what features of a given letter make it that letter and no

> other. If it were otherwise, the simulation of letter recognition would pose no serious problems of a conceptual nature. This points to the conclusion that there is no set of shape or topological characteristics of a given letter which might serve to distinguish definitely between inscriptions of that letter or other letters or marks which represent no letter at all.[6]

Sayre defines "invariant" as "the characteristic or set of characteristics which distinguishes a given class of individuals from all other classes, and the possession of which qualifies an individual for membership in that class."

With respect to Sayre's first point, I would only emphasize that nothing very important follows from the fact, if it is a fact, that human beings are able to recognize items as belonging to a certain type without explicit awareness of an invariant shape or topological feature or set of such features. For this would leave open the question of whether human beings are in some sense implicitly aware of such invariants and rely upon them in their pattern-recognition competences. This is not to suggest, however, that invariant shape and topological features are made use of; it is only intended as an objection to the supposition that if they are, they must, at the moment of utilization, be available to our consciousness. (Compare: The sense in which a child has been said to know—recognize—sentences as grammatical without being able to articulate the criteria by which he does this.)

But the problem we sidle up to in all this is *the* central problem of pattern-recognition. That is, the problem that arises when one actually tries to enumerate characteristics (implicitly operative or otherwise) of objects belonging to a certain class in a way that permits variation and novelty among members within that class and at the same

[6] Sayre, *op. cit.*, p. 98.

time excludes obvious non-members from that class. For example, it is natural to presume that an adequate account of our ability to recognize the written character "M" would explain why and how certain characteristics of M's enable us to see both m and M as M's. What Sayre believes is that there will be no shape or topological characteristic(s) that are both necessary and sufficient for a letter to belong to a certain class. In the case, say, of the letter "A" there will be no single visual feature or small set of visual features, such as an apex or bar position, that will be common to all instances of A's. Sometimes the A may lack an apex, sometimes the bar (e.g., in lower-case a's). Sometimes the bar may be close to the top of the letter, sometimes not, etc.

At this impasse it was once tempting to suggest that one might enumerate various (partially overlapping) sets of shapes and topological features such that although no member of any set or set itself would be necessary, each set would be logically sufficient for determining whether a given inscription is a token of a certain letter-type. If this were so it might then seem possible to specify an invariant feature of A's in terms of a long (potentially infinite) disjunction of varying sets of sufficient conditions. The invariant feature of A's then would simply be that they possessed characteristics depicted by at least one of the disjuncts (C_1) v (C_2) v (C_3) .... (C_n).

It is now old hat that in his *Philosophical Investigations* Wittgenstein cautioned us against just such a move. Yet it is still helpful to recount what this involved, for it is not always realized that simply to be cautioned against trying to solve a problem in a certain way is not, in itself, to solve the problem.

Wittgenstein claimed that we classify things or objects together not on the basis of noticing that the objects of a similar kind satisfy a certain fixed set of necessary and sufficient conditions, but on the basis of perceived similarities among them for which he coined the phrase "family resemblances." He asked what it is, for example, for something to be a game. And told us to reflect on the fantastic variety of things we range together under the description "is a game." He wrote:

> 67. I can think of no better expression to characterize these similarities than "family resemblances." For the various resemblances between members of a family, build, features, color of eyes, gait, temperament, etc., etc., overlap and crisscross in the same way.–And I shall say "games" form a family.

He goes on to provide a further illustration of why we would call something (in this case a number) a certain kind of thing rather than another. He asks:

> Why do we call something a "number"? Well, perhaps because it has a–direct–relationship with several things that have hitherto been called numbers; and this can be said to give it an indirect relationship to other things we call the same name. And we extend our concept of numbers as in spinning a thread we twist fibre on fibre. And the strength of the thread does not reside in the fact that some one fibre runs through the whole length, but in the overlapping of many fibres. . . .
>
> But if someone wished to say: "There is something common to all these constructions–namely the disjunction of all their common properties"–I should reply: "Now you are only playing with words. One might as well say: Something runs through the whole thread–namely the continuous overlapping of those fibres."[7]

It should first be noticed, however, that quite apart from Wittgenstein's remarks, the major difficulty with any attempt to account for a plurality of objects' being the same kind of object in terms of their satisfying any one of

[7] *Philosophical Investigations,* p. 32e.

the varying finite sets of conditions that collectively form a potentially infinite disjunction is that the disjunction will eventually embrace an infinite number of disjoint sets of conditions so that the description of the *kind* in question will be associated with infinitely many utterly different sets of conditions. For example, suppose the disjunction consists in an infinite number of four member sets of conditions thus:

$$C_1[a \cdot b \cdot c \cdot d] \lor C_2[b \cdot c \cdot d \cdot e] \lor C_3[a \cdot c \cdot d \cdot e] \ldots\ldots\ldots C_n[-\cdot-\cdot-\cdot-]$$

Here we have a case where a description, call it "D," applies to an item whenever it satisfies any of the sets of conditions $[C_1] \lor [C_2] \lor [C_3] \ldots \lor \ldots [C_n]$. But although we can see why D would apply to both members of a pair of objects where object O_1 satisfied C_1 and object O_2 satisfied C_2, it is obvious that where the disjunction is infinite (or somewhat large) D will also be associated with some C_i, where C_i is disjoint with at least one and quite possibly indefinitely many other set(s) (C_1, C_2, etc.) with which D is associated. But no account of *kinds* that entails that the description expressing the kind in question is ambiguous can be remotely satisfying. That is to say, where a "kind" description ("is a game," say) can be applied to an object whenever it satisfies either of even two disjoint sets of conditions, we have either a case of a radically ambiguous word or, worse, two different words with the same sound and spelling. But any attempt to formulate rules of pattern-recognition which systematically leads to such a radical ambiguity in the description of the class in question or to a proliferation of words similar only in sound and spelling is obviously foreordained to fail.

To return to Wittgenstein's remarks, however, it is hard to suppress the uneasy feeling that they leave us playing with metaphors. Instead of an alternative theory to replace the discredited one, we are told that we group things under a common description (say, "is a game" or "is a number") on the basis of family resemblances. But how is this really very different from saying we do in fact recognize a variety of things or objects as being of a certain sort on the basis of their resemblance one to another? To point this out is not to provide an account or explanation of anything; it is simply to point to that which is in need of an account or explanation. It is in effect to point to the need of a theory of pattern-recognition.

For this reason I find it curious that Hubert Dreyfus, in his robot-baiting RAND monograph numbered P-3244 "Alchemy and Artificial Intelligence," should cite Wittgenstein's doctrine as a kind of counter-example to current pattern-recognition programs. He seems to treat the notion of "family resemblances" as if it were hooked to an obvious fact of perceptual experience, and as if by mentioning this fact we were treated to some sort of understanding of an unmechanizable phenomenon. But surely the notion is a problematic and semitechnical one to be introduced, if at all, as an acknowledgment that we have a competence for generalization that we do not yet begin to understand. Such acknowledgment hardly qualifies as explanation.

It is little advance over previous theories to be told, as Dreyfus tells us, that

> Recognition of a member of a "family" is made possible not by a list of traits, but by seeing the case in question in terms of its similarity to a paradigm (i.e. a typical case) . . . (p. 44)

for we are not told what "seeing the case in question in terms of its similarity to a paradigm" amounts to. There is, so far as I know, only one clear construal of "recognition" based on use of a "paradigm." But this, ironically, would be wholly alien to the sort of explanation that I know would satisfy Dreyfus. The construal, which I do not attribute to Dreyfus, amounts to viewing "use of a paradigm" as a template-matching procedure whereby the letter "A," say, relative to some grid is digitized and used as a paradigm to be superimposed on whatever character is presented to the machine for consideration. A certain degree of "fit" will then be necessary (and sufficient) for the machine to output "is an 'A.'" Obviously this only brings us back to another disconcerting encounter with an insufficient and unnecessary set of allegedly necessary and sufficient conditions the inadequacy of which the notion of "family resemblances" was introduced to remedy. That such a template procedure is deficient can be seen in the manner by which letters smaller or larger than the template would be easily missed. For example, consider

where either the shaded A or the unshaded A could be viewed as the template and the other as the character-token it fails to "recognize" as an A.

There exists a variety of ways in which the sample may be magnified or reduced, etc., and hence made more susceptible to the template. But as Selfridge and Neisser point out ("Pattern Recognition by Machine")

> . . . Such a procedure is still inadequate. What it does is to compare shapes rather successfully. But letters are a good deal more than mere shapes. Even when a sample has been converted to standard size, position and orientation,

it may match a wrong template more closely than it matches the right one.[8]

In brief, I find Dreyfus' notion of "similarity to a paradigm" either intelligible, (construed as "matching a template") but clearly defective as an explanation of recognition competence or as mysterious as recognition competence itself. (Compare: Noam Chomsky's celebrated criticism of the Skinnerian use of the notion "stimulus generalization.")

Wittgenstein's underdeveloped doctrine of family resemblances leaves us bereft of any account of what might be called "perceptual projection"—the ability that human beings have to recognize, on the basis of a finite number of experiences, a potentially infinite number of new and varied objects they have never encountered before as all being of a certain specified kind.

Here I should like to extend, without mishap, I believe, the use that the word "projection" has in Chomsky's work in grammar and Ziff's and Katz and Fodor's (and Katz and Postal's) work in semantics.

What these writers call "the projection problem" is roughly synonymous with the problem of formulating the rules underlying what may be conveniently viewed as our syntactic–semantic pattern-recognition competence. I tend to view pattern-recognition (and production) as the general case and regard any phenomenon involving learning or skill as special instances of it.

And I think that even rather monorail skills such as learning to ride a bicycle can be seen as comprised of projective capacities, i.e., capacities that enable us on the basis of an exposure to a finite number of trials on a finite number of bicycles to acquire the ability to ride a potentially infinite variety of bicycles in an infinite variety

[8] In *Computers and Thought,* ed. Edward A. Feigenbaum and Julian Feldman (New York, 1963), p. 243.

of ways ("Look, Ma, 28-inch wheels and no hands!" etc.). Were this hunch to prove accurate, it would probably have the side effect of bringing Ryle's cases of "knowing how" under the same general descriptions as his cases of "knowing that."

Very roughly, wherever we find a phenomenon that requires for our comprehension of it the formulation of and solution to a projection problem, the problem of "how to go on," we have a phenomenon involving pattern-recognition, and vice versa.

Wittgenstein, of course, was not consciously addressing himself to problems in pattern-recognition. He was, in effect, I believe, attempting to answer questions left unresolved after years of philosophical haggling over the nature of so-called "real definitions" (the essence or defining properties of given things or substances). The interesting aspect of this cluster of problems is that what is wrapped up in any solution to them seems roughly equivalent to whatever it is which would constitute a solution to recognition problems. In other words, if we were able to explain how it is we come to recognize something as being a certain sort of thing, we would also be able to explain what it is for a certain sort of thing to belong to one class of objects rather than another. In this connection, it is hard to decide if one should rejoice in the potential "psychologizing away" of a metaphysical problem or despair in the expectation that problems analogous to the hoary metaphysical ones will reassert themselves in the context of simulation psychology.

To return to those remarks of Sayre's quoted at the beginning of this section, however, I would agree that it seems highly unlikely that any powerful *simulation* of human pattern-recognition capacities could be devised primarily in terms of rules that were sensitive only to shape and topological features.

Though obvious to some, it may be useful to recount just why reliance on such features is so unsatisfactory. This is so for the unsimple reason that human recognition does occur within a context that very often plays a crucial role in determining how something is recognized as being a certain sort of thing. So it seems strategically necessary to take such features into account at the very outset.

For example, given the second letter in CATS AND DOGS, it hardly matters whether the apex is closed, since the context will disperse all doubt as to whether an open apex A were an A or not an A. Here it is obvious that our ability to recognize an ambiguous inscription as being an A and not an H is partially determined by context. The importance of this example is not that it shows that the closed apex is not a necessary feature of A's, but that it shows that contextual features enter into the set of conditions sufficient for distinguishing A's from other letters in general. More emphatically, trying to decide whether an H-A-shaped inscription with either an open apex or converging sides is an A or an H in isolation is obviously an idle decision, since there is clearly no right choice to be made. Such decisions would have the flavor of real choices only if made against a background of real alternatives. In isolation an H-A-shaped inscription that could be construed as having either an open apex or merging sides *is neither* an A nor an H; it is simply an H-A-shaped inscription. To ask, "But which, really, is it?" is like asking what the "deep structure" analysis is of a token of "Flying planes can be dangerous" when that token is produced by a parrot or found written by the wind in the sand. So too for the following inscription:

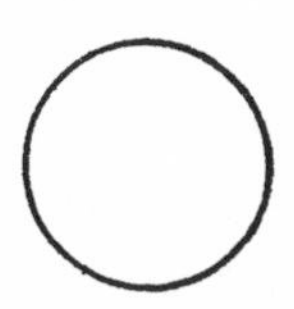

Is it an "o" as in "cod" or a zero as in $1.00? How could we decide to opt for either of these descriptions? Why not say it is O-shaped or zero-shaped, or shaped *like* an o or shaped *like* a zero? Suppose a cloud drifts over our head in that shape. Is it the letter "O" written in fluff? (Here I shall brush aside the clever rejoinder that what it really is is an inscription for illustrating the inappropriateness of the foregoing questions.)

Compare: A farmer might use a rock from his field for a doorstop and we might recognize it as such on the basis of how it looked. But imagine trying to decide whether a given rock is a doorstop apart from the context in which it is one. ("See this rock?" [pointing to a rock on the coffee table], "Well is it or isn't it my doorstop?")

But once the contextual determinants are brought into the open—not that they had to be—there they are, always leering over our shoulder—it seems plain that the recognition of even simple characters will (often) depend on other levels of recognition abilities such as syntactic and semantic ones. For example, if we recognize the second to the last letter in the sentence "The cow chews cud" as an almost closed u and not as a somewhat open o, it will no doubt be because of what we know about the meaning of the word "cud" and the meaning of the word "cod" (as well as what we know about the eating habits of cows). If successful "disambiguation" of letter script generally depends on, say, semantic abilities, then it would be unreasonable to seek a general explanation of letter-recognition ability distinct from an over-all theory of language-learning capacities. This should be emphasized since recognition programs for individual letters have sometimes been discussed as if they were of interest apart from a general theory of linguistic capacities.

It may be distressing as well as exhilarating to a character-recognition theorist to find that problems in semantic interpretation crop up in connection with the recognition of seemingly simple characters. It is exhilarating in that the problem of constructing simple character-recognition programs turns out to be a much deeper problem than it was at first believed to be. For it may at first have seemed that letters were a kind of minimal linguistic unit—building blocks of words, etc.—and that it would be easiest to devise a recognition program for them and then, hopefully, build up from there. What we soon find, however, is that recognition or interpretation at any one level of linguistic description (letter, phonemes, word, sentence, discourse) is a partial function of how we can construe the units at other levels. Or, as I would prefer to put it: disambiguation problems occur for every level of linguistic description and, in fact, for levels of non-linguistic behavior as well. (Compare: Having observed a man's movements we ask, "Is he in pain or only pretending?") There is, thus, no "easiest level" as such for which to devise a pattern-recognition program.

Unfortunately, attempts to mechanize the process by which disambiguation takes place have been notoriously unsuccessful. As A. Oettinger has remarked, "The major problem of selecting an appropriate correspondent for a Source Word on the basis of context remains unsolved, as does the related one of establishing a unique syntactic structure for a sentence that human readers find unambiguous."

The major obstacle is that there is still no precise account of how one could build into a program the fantastically varied background knowledge (e.g., that cows don't eat fish) that human speaker-hearers bring to a communication situation and that enables them in fairly systematic and facile ways to interpret correctly each other's utterances, even though considerable novelty may attend both the syntactic, and situational or pragmatic features of the remark that is interpreted.

Nevertheless, the problem of how to incorporate sensitivity of context into a mechanical disambiguater or a conversational robot seems fairly well defined. It is obviously not enough to say that the ordinary speaker-hearer interprets or disambiguates utterances on the basis of context, for this leaves open how in turn he perceives that this or that is a contextual feature that contributes to a correct interpretation of the utterance in question. If one says he does the latter on the basis of *further* contextual information, we are faced with a regress (i.e., we need to know that such-and-such determines that such-and-such constitutes the context by which we know that such-and-such is the context by which we interpret a certain utterance.) But since human beings do in fact disambiguate utterances and, in doing this, do in fact rely on contextual information, and since they could not so rely on such information if such reliance lead to a regress, clearly in practice there is no regress, though of course there are mistakes. Part of the problem for the psychologist, then, or the computer-simulation theorist is to recapture, in programming language, just how such a regress is avoided or how one generally limits his search and "cuts off" a potentially ever-expanding intake of *possibly* relevant contextual factors and at a convenient point makes a successful stab at interpretation. I would suppose that in some manner or other a portion of sound or space *becomes* a context only when a human act, verbal or otherwise, or something analogous to such an act occurs there. That is, part of what must be accounted for is this. The meanings of inscriptions (etc.) are not only a function of the contexts in which they occur; that something is a context is in part a function of the act taking place there. More graphically put, finding a message in a bottle not only changes how you see the message, but also changes how you see the bottle. Or, not only do contexts help to determine meanings, but meanings also help to determine contexts.

I would also suppose that the degree of contiguity required between a context and an act subject to interpretation is in part a function of the current "perceptual scope" of any speaker-hearer (actor-interpreter). For example, when someone in Minneapolis cannot, as he usually cannot, see, and assumes no one else can see, what is happening in Bombay (or even next door, usually), then when a speaker-hearer in Minneapolis says something to another speaker-hearer in Minneapolis, it is assumed (generally) that what is happening in Bombay is not part of the context in which the utterance uttered in Minneapolis occurs. The point is that most of the time such an assumption would be very useful to make, just as in chess it is usually, though not always, useful to protect one's queen. That is, such an assumption probably functions as a strategy or heuristic or rule of thumb, one of many, that human beings use in attempting to understand each other. There is, surely, no algorithm or "fail-safe" technique by which human beings communicate with each other. (A succinct way of paraphrasing "to err is human . . ." may be this: Infinitely expanding contexts force "cut offs," meaning the use of non-"fail-safe" heuristics and this necessitates the possibility of error.) A reconstruction of whatever the coalition of heuristics amounts to that human beings use constantly, quite unconsciously and with utter ease, to make themselves understood and to understand others would amount to a theory of contexts (and vice versa). That no one, I believe, at the moment knows in any detail what such a reconstruction would look like is important and interesting. But it should not be stressed to the point of profligate pessimism. There is no reason to suppose, as, for example, Hubert Dreyfus seems to suppose, that current difficulties in formulating a theory of contextual sensitivity that might be simulated on a computer can be converted into an argument that claims that it is impossible to simulate by computer the sensitivity to contexts ordinarily exhibited by human beings. (See Dreyfus' *Critique of Artificial Intelligence.*)

Yet even Benjamin White ("Studies of Perception"), who is otherwise optimistic, writes:

> None of these computer recognition programmes, for example, has tackled the problem of segmentation. In every case, it is assumed that there is but one letter, one vowel, one pattern in the computer to be recognized. No attempt is made to isolate the spoken vowel in a stream of connected speech, not to isolate a single letter in cursive script.[9]

That disambiguation occurs at all linguistic levels and that human beings systematically "bring it off" at all levels indicate that a "projection problem" could be formulated for each level. Any fully integrated theory of linguistic and non-linguistic behavioral descriptions would attempt to formulate projection rules for each level of competence and explain each level's interactions with the others.

A few further remarks should be made concerning the nature of the projection problem in general. Consider again the case of letter-recognition. Even with a template procedure, the machine may "project" to new cases. There may be indefinitely many novel A's that vary stylistically from the template letter but which the matching procedure will recognize as A's. For example, suppose the letter under consideration varies only from the template in its two sides being much longer. Then the template would match perfectly with the upper part of the input A. Such a degree of match we can imagine as being sufficient for the output "is an A," even though the letter being recognized as an A differs from the template A in

[9] In H. Borko, ed., *Computer Applications in the Behavioral Sciences* (Englewood Cliffs, N.J., 1962), p. 298.

that the sides are much longer. One can even imagine indefinitely many novel (non-template) A's which would conform to instances of variations on this simple difference in length of sides. (Compare: One might formulate a definition of "grammatical sentence in English" such that if a sentence was of the form NP + V + NP, it would be counted as grammatical. Thus "The man + hit + the ball" would be grammatical; "the girl + hit + the ball" would be grammatical; "the penguin + ate + the field linguist" would be grammatical; and so on. Now use of such a crude "syntactic template" would not do justice to our over-all intuitions concerning grammatical sentences in English; yet it would account for indefinitely many novel English sentences. The template can be seen as the embodiment of a rule that would enable us to encode or decode indefinitely many morphemically novel grammatical sentences, but it is not a rule that would result in *relevant* projection to an infinite set of novel sentences. (By "relevant" here I mean relevant in that it would capture the scope of what is intuitively meant by "grammatical sentence in English.")

So, too, the problem with a template-matching character-recognition program is not that it necessarily fails to provide for a potential infinity of variations, but rather that the variations involved are slight, so the way in which the match is performed is obviously exceedingly crude and inflexible compared to however it is that human beings do this.

This problem can be clarified by sorting out the following very different goals that a pattern-recognition program (designed to recognize the letter M, say) might be expected to attain if it is to reflect human abilities. Let us assume the machine and the program are in good working order and let us assume that the machine is to make a decision "M" or "~M" whenever a symbol is presented to it in a certain manner. Then,

(1) the pattern-recognition program might be deemed successful if and only if the machine outputs "M" when the letter M (in some stylistic form) is presented to the machine under suitable circumstances.

Or,

(2) the pattern-recognition program might be deemed successful if the machine outputs "M" when the letter M (in some stylistic form or another) is presented to the machine under suitable circumstances.

Or,

(3) the pattern recognition program might be deemed successful only if the machine outputs "M" when the letter M (in some stylistic form or another) is presented to the machine under suitable circumstances.

Or,

(4) the pattern recognition program might be deemed successful if the machine generally outputs "M" when the letter M (in some stylistic form or another) is presented to the machine and generally fails to output "M" when some symbol other than "M" is presented to the machine under suitable circumstances.

Now if we assume that the point of a given pattern-recognition program is to simulate human recognition abilities and processes, it seems clear that (1) would be an unreasonable goal. For first, it is obvious that even intelligent human beings fail on occasion to recognize a written letter for the letter that it is. And second, not only is it reasonable to suppose that human beings will on occasion fail to recognize a letter, say M, as an M, but will on other occasions think that they claim to recognize as an M what is really a non-M. Hence (2) turns out to be unreasonable as well, given the first cited reason for rejecting (1) as a goal. And (3) hence turns out to be unreasonable for the second reason cited in rejection of (1). This, unfortunately, leaves (4). I say "unfortunately" simply because (4) as stated above is a decidedly less well-defined goal than either (1), (2), or (3). This is so since what is meant by "generally" will have to be relativized to a certain relevant set of letter presentations. That is, it is possible to imagine a set of character presentations that are especially difficult for the machine or a human being to recognize as being this or that letter, so that although the machine (or human) generally failed to recognize an M as an M, we would not count this as showing that the machine (or human) did not possess capacities for the recognition of M's. (Compare: A human being may be unable to understand what another human being says because the latter speaks so rapidly. This does not show that the human being is incapable of understanding spoken utterances, but only a certain idiosyncratically produced subset of them.) What would be needed to carry through on goal (4) would be a definition of "relevant set of character presentations." I do not wish to go into the problems confronting the formulation of such a definition for a particular character, but should like only to record my suspicion that I find it highly implausible if such a definition might be forthcoming for a single character. One might try to form a definition by relativizing "generally outputs 'M' " to "when the letter M presented to the machine assumes a standard form." But what, then, would be meant by "standard form"? What, for example, would count as the standard size, shape, style, etc. of the letter M? This seems to me to be an obviously frustrating inquiry. (Is the size, say, of a billboard M less standard than this one: m?)

V

Turing's Test and the Need for Evaluation Procedures

It is indicative of an important characteristic of CS research that Sayre should assume, without argument, that the goals of most pattern-recognition studies have been to devise models or simulations of human cognitive processes. He writes:

> Let us imagine that the system has been improved to give 95% correct classifications over a series of experiments involving, for example, one dozen writers. Should we consider that the attempt to simulate human letter-recognition now is almost completely successful? What we would be justified in thinking is that the machine produced classifications of several letters which agreed in 19 out of 20 cases with the identifications provided by the dozen people who wrote them. It undoubtedly would be an accomplishment to build a machine which could do this. But we are free to ask exactly what relevance such a machine would have to the task of building a mechanical system which performs comparably to human beings in the recognition of letter-inscriptions.[10]

Now although the stated aims of pattern-recognition research have been somewhat equivocal, such work should not be characterized as having been primarily directed toward providing a *simulation* of human pattern-recognition capacities. To a large extent it has been oriented toward the goal of producing machines that perform certain tasks whether or not its procedures in performing those tasks mirror human procedures for executing similar tasks. Pattern-recognition research has, in other words, been research in artificial intelligence with

[10] Sayre, *op. cit.*, p. 205.

the emphasis on "artificial." Hence, given that the goals of pattern-recognition have not been conceived by researchers to involve replication or even a looser simulation of human cognitive processes, the dangers inherent in any attempt to judge the success of pattern-recognition by standards appropriate to CS should be apparent.

I would not bother with "setting the record straight" were it not the case that there had been a *systematic* confusion between the goals of CS and the goals of AI. But even this would be without philosophical interest were it not symptomatic of something else, namely, the lack of any compelling evaluation procedures for deciding between *competing* simulations of human cognitive processes.

That Sayre was not alone in this confusion, and that it has been indicative of an interesting problem given the current state of the art, can be made clearer by examining Hao Wang's reactions to some of the CS work of Simon, Shaw, and Newell.

Marvin Minsky's excellent article "Steps Toward Artificial Intelligence" contains a criticism of Hao Wang's assessment of Simon, Shaw, and Newell's investigations with their program called the "Logic Theorist" (LT), which they used to prove theorems in *Principia Mathematica.* Minsky claims that Wang

> . . . has criticised the LT project on the grounds that there exist, as he and others have shown, mechanized proof procedures which, for the particular run of problems considered, use far less machine effort than does LT and which have the advantage that they will ultimately find the proof for any provable problem.[11]

Minsky (rightly) points out that Wang misconstrued the intentions underlying the LT project. It was not supposed to be an exercise in effective proof procedures, but was

[11] In Feigenbaum and Feldman, *op. cit.,* p. 437.

designed instead as an attempt to utilize so-called "heuristics" (non-"fail-safe" strategies) with the end in mind being a program yielding a TRACE that constituted a simulation of human problem-solving processes. Just as Sayre mistook work in AI as being directed at the goals of CS, Wang has mistook work in CS as aiming at the goals of AI. Thus Wang writes:

> They do not wish to use standard algorithms such as the method of truth tables, because "these processes do not produce a proof in the meaning of Whitehead and Russell. One can invent "automatic" procedures for producing proofs, and we will look at one briefly later, but these turn out to require computing times of the orders of thousands of years for the proof of *2.45." [The embedded quote is from "Empirical Explorations of the Logic Theory Machine."] It is, however, hard to see why the proof of *2.45 produced by the algorithms to be described in this paper is less acceptable as a proof, yet the computing time for proving *2.45 is less than ¼ second by this algorithm. To argue the superiority of "heuristic" over algorithmic methods by choosing the particularly inefficient algorithm seems hardly just.[12]

But Simon, Shaw, and Newell are not interested, as Wang believes, "in seeing how well a particular procedure can enable us to prove theorems on a machine." Their work in theorem-proving is of a piece with their over-all efforts to simulate human cognitive processes. In general these processes do not lend themselves to the employment of efficient algorithms such as those adopted by Wang. This is clearly so in the case of chess-playing or so-called "creative thinking"–music composition and the like. Consequently, the fact that Wang could devise a better algorithm for the set of problems dealt with by LT is neither here nor there so far as Simon, Shaw, and

[12] Hao Wang, "Toward Mechanical Mathematics," *IBM Journal of Research and Development,* 1960, 4, pp. 2–22.

Newell's central purpose is concerned. That they chose to deal with a set of problems for which efficient algorithms exist, and which human problem-solvers such as Wang might adopt, may be unfortunate. Such theorem-proving obviously does not best represent the kind of human problem-solving they are interested in simulating. But that is another, and minor, matter.

Nevertheless, sympathy is due both Sayre and Wang with respect to their misunderstandings. For these turn out to be remarkably understandable misunderstandings once it is realized how a considerable number of researchers in CS, including from time to time Simon, Shaw, and Newell, have tended, albeit with some misgivings, to appraise their own work.

For any given simulation of human cognitive processes, the question arises as to whether the simulation is an accurate or "good" simulation. Virtually the only evaluation procedure in consistent use by researchers in CS is what has been called "Turing's Test."

Two adaptations of Turing's "Imitation Game" have found their way into the literature of CS (as well as AI). The first is equivalent to Turing's version (hereafter Turing's $Test_1$) which was intended as a specific single test, the passing of which was construed as *a paradigm* of intelligent behavior (or given its use in CS–a paradigm of a simulation of intelligent behavior). The second is a much weaker and, unfortunately, more widely used version of the first, it consists simply of comparing human and machine outputs (or end results) for *any* given task which hitherto required execution by an intelligent agent. If the outputs, taken by themselves, seem indistinguishable, then intelligence is attributed to the machine (at least to some extent) and the simulation is regarded as successful. We shall call the latter Turing's $Test_2$. Turing, however, never explicitly intended his test to be one which was to be applied skill by skill, situation after situation; he was interested in a *general* test of intelligence, i.e., the Imitation Game. The question of whether the Imitation Game is such a test, we have already discussed and answered in the negative. But in fairness to Turing, he certainly need not accept *any* intelligently executed human task that a machine might execute as well as a human being as equivalent to success in playing the Imitation Game as well as a human being. So Turing's $Test_2$ is at least *prima facie* a different version of a test for (simulated or literal) thought and intelligence from Turing's in that it relativizes it to any particular task.

It might be suggested that research in CS has not proceeded under the assumption that any instance of Turing's Test's being passed would provide an argument for the machine's being intelligent or providing a good simulation of intelligence. Instead it could be claimed that passing the test would be a weak way of showing that the machine is, *for that situation,* as intelligent as a human being or a good simulation of such intelligence. What this overlooks, of course, is that first of all it might be possible to enable a machine to pass the weaker version of Turing's Test for a wide range of situations and still not thereby enable it to play well Turing's Imitation Game. And second of all, how we appraise the closeness of the analogy for any given situation should depend in part on how we appraise the analogy over-all. The defect in such simulation strategy is interestingly enough the same defect there was in Montaigne's argument (in his "Apology for Raimond Sebond") for animal intelligence that was discussed in Chapter One. There we noted that Montaigne had argued that *animals in general* were as intelligent as human beings because, for example, the fox could do such-and-such a thing that if done by a human being would be regarded as intelligent, the tunnies could do such-and-such a further thing that if done by human beings would be regarded as intelligent, and so on. What Montaigne failed to see, and what CS theorists have similarly failed to see, is that even if human intelligence could be specified in terms of intelligent tasks, it is simply invalid to argue that if for every intelligent human task there were an animal (or machine) that could perform that task, then animals (or machines) would be as intelligent as (or would at least simulate the intelligence of) human beings. For this would be like arguing that if I have a bag of marbles with twenty-five different colored marbles, then if there are twenty-five other bags each with one marble the color of one of my marbles, then there are other bags of marbles that match mine with respect to variety in color.

And furthermore, it does not seem possible to specify human intelligence simply in terms of intelligent tasks.

The general philosophical weakness of Turing's Tests as stated by Turing was elaborated in Chapter Two. Those criticisms can also be applied to the use of Turing's Test_1 and *a fortiori* to Turing's Test_2 within the context of CS. Suppose, for example, that one programmes a computer using IPL-V to solve the simple problem of scanning a list to see if a certain symbol is on it and, if it is, to delete it. (In IPL-V such a routine, call it R1, would be constructed primarily from two basic processes: J77 which is a search-process for specified symbols and J69 which is a deletion instruction.) Now suppose one asks whether R1 provides a "good" simulation of a human being who searches a list for a symbol, finds it, and deletes it. Call the hypothetical human routine R1′. How do we tell if R1 is a good simulation of R1′? If we use Turing's Test_2, then whenever we find the list-as-deleted by R1 and the list-as-deleted by R1′ are indistinguishable, R1 will qualify as a good simulation of R1′. But imagine the different kinds of ways that could be typical of how a human solves such a problem: If the list is short enough the human might scan it all at once, center in, and then delete; or he might scan the list sequentially item by item, then delete, etc. These ways, however, could hardly be expected to exhibit themselves on the list with deleted symbol. Thus comparable end results, a list with deleted symbol, or what is tantamount to passing Turing's $Test_2$, can hardly be sufficient to guarantee that the information processes depicted by the computer program TRACE are the same as, or even usefully analogous to, the psychological or psychophysiological processes characteristic of human beings achieving the result in question. Furthermore, and this returns us to Sayre and Wang's confusion, to rely on Turing's $Test_2$ in CS research is to abrogate the very real distinction in goals between CS and AI. For the goals of AI are, in effect, defined as programming a machine to pass Turing's $Test_2$–i.e., to program a machine to bring about end results that hitherto required intelligence as cause. *To invoke Turing's $Test_2$ as the primary evaluation procedure for programs in CS suggests that CS is equivalent to AI–for the criteria for achievement at either are identical. This is absurd. It would make it logically impossible to simulate an end result without thereby simulating the process by which that end result was achieved!*

If, however, we take the real aims of CS seriously–the goal being to simulate psychological *processes*–then it should be possible for R1, say, to be an excellent simulation of R1′ even though there is dramatic *mis*matching at the task level. Another way of putting it: If there is to be any psychological import for simulation studies, Simon, Shaw, and Newell should be willing, in principle at least, to claim success for LT or a program like it *even if it fails to prove a single theorem!* For excellent simulation could still obtain at the process level which is, supposedly, what CS is designed to simulate.

Yet time and again the claim that a successful simulation has been achieved is based solely on a program's having passed Turing's $Test_2$—or having succeeded in satisfying the goals of AI. And for the most part, *only* programs that have some sort of dramatic end results to speak for them are publicized. So it is little wonder that someone such as Sayre or Wang should conflate the two sorts of ventures. Wang has, in a sense, devised a program that enables a machine to pass Turing's $Test_2$ with a vengeance. (It might be argued that passing the test "with a vengeance" is not passing the test.) But if success at CS is equated with passing Turing's $Test_2$, then Wang should be able to claim success at CS for his program.

The mental processes of morons and people of normal intelligence may be psychologically similar in a large number of respects, yet their respective achievement levels (end results) may be dissimilar in almost all respects. Certainly a moron's mind is much more like that of a normal person's than is the information processing system of a digital computer, even though the latter may be much more likely to pass a string of Turing $Tests_2$. (Cf. L. Uhr and C. Vossler's remarks on programs that perform at a higher level than human subjects in "A Pattern-Recognition Program that Generates, Evaluates, and Adjusts Its Own Operators.")[13]

Some of the above objections to the reliance on Turing's Test(s) in CS theory are in essential agreement with and others are in conflict with various remarks made by Julian Feldman in his article, "Computer Simulation of Cognitive Processes." Feldman writes:

> Turing's Test lacks the rigor that the computer simulation of cognitive processes promises to restore to the study of cognitive processes. It is certainly true that the use of judges is the respectable test, and statistical techniques are available for determining the ability of judges to discriminate beyond the mere guessing period. The use of Turing's

[13] In Feigenbaum and Feldman, *op. cit.*, pp. 251–68.

Test, however, implies that no good operational criteria exist for comparing human protocols and machine protocols. Furthermore, the ability of the computer simulation of cognitive processes to achieve its goal–the creation of a model that will produce a reasonable facsimile of the subject's behavior–is probably not very well tested by Turing's Test. Turing's Test is adequate for telling the researcher whether or not he has achieved a program that will produce behavior that is indistinguishable from human behavior. But Turing's Test is not adequate for matching the protocol of the machine's behavior and the subject's behavior; since even though the protocols may be quite different, judges cannot determine which is the machine's protocol. In statistical terminology, Turing's Test is not very powerful for testing simulation models. The probability of an error of the second kind, i.e., accepting the model when it is in fact wrong, is quite high.[14]

Feldman's attempt to find a rigorous alternative to Turing's Test has led him to employ the technique known as "conditional prediction." This he describes as follows:

> Every time the program makes a decision different from that of the subject, the program can be set back on the track, i.e., the subject's decisions imposed on the model, so that the differences between the subject's behavior and the program's behavior at any point in time will not be caused by the differences that occurred earlier. Thus the decisions of the model at any point in time are conditional on the preceding decisions of the models being the same as those of the subject.[15]

Conditional prediction is an interesting tactic because at first peek it seems to exhibit alternative criteria to Turing's Test(s). Further peering, however, leads one to ask how conditional prediction leads one beyond Simon, Shaw, and Newell's application of Turing's Test

[14] In Borko, *op. cit.*, pp. 347–48.
[15] *Loc. cit.*

"all along the line"—i.e., at each decision juncture. What is important to notice here is that given the theoretical situation depicted in diagram #1 of this chapter (p. 97), conditional prediction simply *is* Turing's Test relative to deeper levels of comparison (say, levels 2 and 3) and hence inadequate as a test to determine successful simulation at those deeper levels. That is to say, even if conditional prediction enabled one to match *behavior* at the protocol level, a matching of behavior at a protocol level cannot by itself guarantee similarity or interesting analogy at deeper levels. But it is precisely at these deeper levels that cognitive processes exist, as distinct from, say, their verbal indicators or protocols.

Feldman is correct in claiming that Turing's Test is inadequate for determining a match between (even!) the machine's protocol or behavioral trace and the protocol (or verbal reports) of a human subject's behavior. But he is incorrect in (tacitly) suggesting that a rigorous alternative to Turing's Test would be a test that *did* decide when such a match had occurred. For a match at the level of protocols is still a match at a very superficial level—superficial, in that protocols should not be identified with the psychological processes giving rise to them.

It would certainly be utopian to try to devise a way of detecting the *best possible* simulation, but it is not utopian, and is indeed minimal, to be able to have some criteria for evaluating competing models. Suppose, for example, Uhr and Vossler's use of operators provided as efficient a program as, say, Frishkopf's or Sayre's own program. It is at this point, I think, that an underlying weakness of most current work in CS comes to the surface. Success at producing programs that perform tasks as well as human beings do only appears efficacious as a criterion so long as there are relatively few successes at the task level. As soon as numerous programs equivalent at task-level performance appear, the shallowness of using similarity in end results as a basis for assessing simulation of the process leading to these results becomes apparent. Obviously all programs are not equally good depictions of human processes, but at present there are few if any evaluation procedures to fall back on to facilitate a choice between competing program simulations.

To follow Chomsky's suggestions with respect to the aims of a theory of grammar, what now appears to be needed is a *theory* of simulation such that given two competing simulations S_1 and S_2 of a mental (or biological) phenomena P, the theory would tell which was the better simulation. In other words:

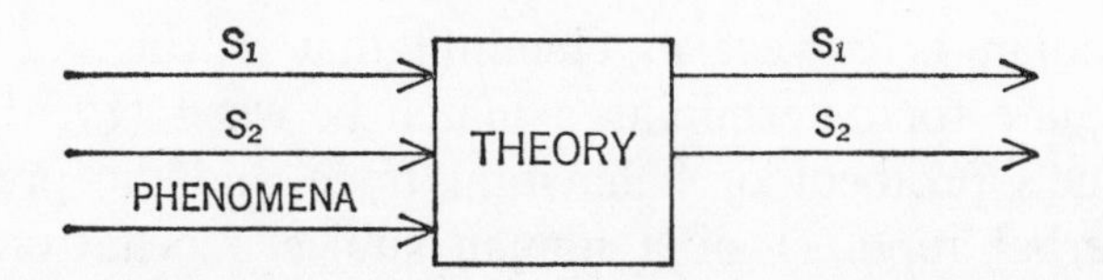

where the "output" of the theory is S_1 or S_2 but not both.

This need for an evaluation procedure is likely to become more acutely felt as an increasing number of simulation programs become roughly comparable in enabling a computer to perform certain tasks. It would indeed seem remarkable if all programs that were equivalent in task-performance results (proving theorems, say) were thereby equivalent as models of human mental processes that lead to similar results.[16]

Turing's Tests have, unfortunately, been a sort of methodological sop offered to those who think there are no criteria whatever for evaluating simulations of psychological processes. But the extent to which they are relied on is the extent to which CS theorists remain content to concern themselves with the surfaces of behavior and ignore the deeper underlying processes that produce it.

[16] Cf. my "Minds and Machines: A Survey," in R. Klibansky, ed., *Contemporary Philosophy: A Survey,* Florence, 1968.

This may be punctuated in a manner that reveals still further problems with current CS tactics by considering the use of "protocols" in writing programs.

IV

Protocols, Introspection, and Behaviorism

A protocol is simply a subject's report of what he is doing while working through a problem presented to him by the CS experimenter. This record then forms the basis upon which the computer program (say, in IPL-V) is constructed. In short, one interrogates the subject and then finds a suitable basic process or complex routine in the programming language to reflect what the subject reports himself as doing. These are strung together in an appropriate fashion and thus an entire program for the problem-solving process is obtained. (The experimenter may, however, have to use his own discretion in substituting routines for what he takes to be gaps in the subject's reporting.)

Newell and Simon in their study, "GPS: A Program That Simulates Human Thought,"[17] provided a human subject with a problem as follows:

$$(R \supset \sim P) \cdot (\sim R \supset Q) \quad | \quad \sim(\sim Q \cdot P)$$

The subject was not told that this is a problem in elementary symbolic logic. He was only informed that he had twelve rules (described for him by the experimenter) for manipulating expressions containing letters connected by "dots" (•), "wedges" (V), "horseshoes" (⊃), and "tildes" (~), which were to stand respectively for "and,"

[17] In Feigenbaum and Feldman, *op. cit.*, pp. 279–93.

"or," "implies," "not." The subject was given practice in applying the rules, but prior to the experiment he had worked on only one other problem of that sort. The experimenter informed him that he should try to obtain the upper right corner expression from the expression in the upper left corner using his twelve rules. At any time the subject was able to request that the experimenter apply one of the rules to an expression that was already on the blackboard. Whenever the transformation was acceptable, the experimenter would write down the new expression in the left-hand column with the name of the rule in the right-hand column beside it. The subject was asked to talk aloud as he worked. The record of this, the protocol for the first part of this particular experiment, is provided below:

> "Well, looking at the left-hand side of the equation, first we want to eliminate one of the sides by using rule 8. It appears too complicated to work with first. Now–no,–no, I can't do that because I will be eliminating either the Q or the P in that total expression. I won't do that at first. Now I'm looking for a way to get rid of the horseshoe inside the two parentheses that appear on the left and right sides of the equation. And I don't see. Yeh. If you can apply rule 6 to both sides of the equations, from there I'm going to see if I can apply rule 7."
>
> Experimenter writes: 2. $(\sim R \vee \sim P) \cdot (R \vee Q)$ "I can almost apply rule 7, but one R needs a tilde, so I'll have to look for another rule. I'm going to see if I can change that R to a tilde R. As a matter of fact, I should have used rule 6 on only the left-hand side of the equation. So, use rule 6, but only on the left-hand side."
>
> Experimenter writes: 3. $(\sim R \vee \sim P) \cdot (\sim R \supset Q)$ "Now I'll apply rule 7 as it is expressed. Both–excuse me, excuse me, it can't be done because of the horseshoe. So–now I'm looking–scanning the rules here for a second and seeing if I can change the R to a not R or a dash R in the

second equation, but I don't see any way of doing it. [Sigh] I'm just sort of lost for a second."

Here the only rule we need concern ourselves with in order to understand the above phase of the subject's behavior is rule 6: $A \supset B \leftrightarrow \sim A \vee B$. In the first instance, the subject applied rule 6 to both the left and right of $(R \supset \sim P) \cdot (\sim R \supset Q)$ and then changed his mind and applied it only to the left. Here, however, I am not interested in the actual character of the subject's derivation, but only in the way in which the simulation of the subject's mental processes was thought by Newell and Simon to be made possible through the use of protocols. The use of protocols is not regarded as a necessary feature of computer simulation studies. Thus, to show that the use of protocols was methodologically unsound would not show that CS studies were unsound *simpliciter*. Nevertheless, willingness to use the protocol method is indicative of certain preconceptions about mentality on the part of researchers working in CS, and it is this preconception that I wish to flush from cover.

The following questions can be asked of the protocol technique: What sort of data do the protocols make available to us? Do we *infer* from the presence of certain protocols that certain thought processes are taking place in the subject? If so, can we then somehow discover that the protocols are accurate indicators of the subject's "cognitive processes," which are, after all, that which is supposedly being simulated? Suppose one answers the just-raised question by claiming that the protocols are not a *basis* from which to infer certain cognitive processes, but are instead identical with the cognitive processes being simulated. As we shall see, this stance toward protocols is sometimes taken. Then, apart from the implausibility of the identification, we are faced with the question of what it is about protocols that needs to be explained. That is, protocols constitute a phenomenon that we generally understand in the sense that simply by understanding the language in the sense of being able to use it we see what a person is up to. We might ask someone, "What are you doing?" and he might reply, "I'm looking to see if this symbol matches any of these." We then say, "Oh I see." And we do! Given an appropriate computer language we can, of course, define a routine that instructs the machine to compare symbols until it finds a match. But what do we gain by this? What further illumination should we expect from constructing a computer program that in some loose sense simply duplicates that behavior? It might be assumed that the "payoff" would come *only* after the processes *underlying* the computer's behavior could be shown to be analogous to the process (as yet unknown) underlying the human's protocol behavior. But this is often not assumed, and there is thought to be a theoretical-explanatory payoff simply in the construction of a program the trace of which is roughly isomorphic to a string of human protocols. It is this that I do not understand.

If, on the other hand, the protocols are not taken to be identical with the cognitive process being simulated, but are taken to be representative of some other "inner" processes that are (possibly) nonverbal, are we not face to face with classical introspectionism and its attendant methodological difficulties? I do not mean to suggest that these difficulties are insurmountable, but only that they should be acknowledged and dealt with.

Julian Feldman ("Computer Simulation of Cognitive Processes," in Borko, *op. cit.*) argues that the problems of introspectionism do not arise. In support of his claim that "the thinking aloud" procedure is not to be confused with introspection, he cites the following passage from K. Duncker's "On Problem Solving" (*Psychological Monographs,* 58, No. 270, 1945):

This instruction, "Think aloud," is not identical with the instruction to introspect which has been common in experiments on thought-processes. While the introspecter makes himself as thinking, the object of his attention, the subject who is thinking aloud remains immediately directed to the problem, so to speak, allowing his activity to become verbal. When someone, while thinking, says to himself, "One ought to see if this isn't——," or, "It would be nice if one could show that——," one would hardly call this introspection; yet in such remarks something is revealed.

Of course "something is revealed." What is revealed is what the person is up to. It is not surprising that a person can often tell us what he is doing, what his plans are, etc. And, of course, the criticisms of introspective methods in psychology could not show, were never meant to show, that we can never really tell what people are up to by asking them. The question remains, what would be gained by simply simulating such tellings with a computer program?

Newell and Simon are not unaware of the introspection issue. They write:

It is now proposed that the protocol . . . constitutes data about human behavior that are to be explained by a psychological theory. But what are we to make of this. Are we back to the introspection of the Wurzburgers? And how are we to extract information from the behavior of a single subject when we have not defined the operational measures we wish to consider?

Their answer to this question is as follows:

There is little difficulty in viewing this situation through behavioristic eyes. The verbal utterances of the subject are as much behavior as would be his arm movements or galvanic skin responses. The subject was not introspecting; he was simply emitting a continuous stream of verbal behavior while solving the problem-solver that explains the salient features of this stream of behavior.[18]

[18] In Feigenbaum and Feldman, *op. cit.* pp. 281–83.

First note that protocols are initially said to constitute data *about* human behavior. Since John B. Watson, "behavior" has been such an elastic sort of term that it is difficult to tell what it is being stretched to include or when it will snap back and sting our fingers. But here, at least, the protocols are viewed as data that are *about* something else. One paragraph later, however, protocols are treated *as* the behavior that is to be studied. This forces the question: "What is meant by 'explain'?" and "What is meant by 'the salient features of this stream of behavior'?" In the above characterization it seems that "the salient features of this stream of behavior" will by definition turn out to be those features of the problem-solving that contribute to the derivation. But once cognitive processes have been assimilated to features of this kind, it is again unclear what is supposed to be explained. The assumed utility of simulating this behavior becomes suspect. The question is, is this what we are to understand by a program that simulates human thought?

A more detailed account of what Newell and Simon consider a good theory of a subject's thinking processes is contained in the following:

> We may then conceive of an intelligent program that manipulates symbols in the same way that our subject does—by taking as inputs the symbolic logic expressions, and producing as outputs a sequence of rule applications that coincides with the subject's. If we observed this program in operation, it would be considering various rules and evaluating various expressions, the same sorts of things we see expressed in the protocol of the subject. If the fit of such a program were close enough to the overt behavior of our human subject—i.e., to the protocol—then it would constitute a good theory of the subject's problem-solving.

And:

> Conceptually the matter is perfectly straightforward. A program prescribes in abstract terms (expressed in some

> programming language) how a set of symbols in a memory is to be transformed through time. It is completely analogous to a set of difference equations that prescribes the transformation of a set of numbers through time. Given enough information about an individual, a program could be written that would describe the symbolic behavior of that individual. Each individual would be described by a different program, and those aspects of human problem-solving that are not idiosyncratic would emerge as the common structure and content of the programs of many individuals.

What I find important in the above passages is that the following conflations have taken place. Cognitive processes become identified with problem-solving behavior, problem-solving behavior becomes identified with verbal protocols, and a set of descriptions of the protocols from which a computer simulation program can be constructed becomes identified with a theory of human thinking in general. Thus, the role of program-writing becomes that of enabling us to make generalizations concerning the sorts of strategies, maneuvers, and so on that human beings make in trying to solve certain kinds of problems. These generalizations are then treated as explanations of human thinking. This needs to be looked into.

First of all, it is not clear how the program-writing provides us with an analytic aid in formulating generalizations about the human behavior. Why not simply generalize about the protocols themselves? But more pointedly, one can ask why we should assimilate the thought processes involved in human problem behavior to the actual behavioral outputs of the human being (here either protocols or jottings of paper [see Fig. 3, p. 95], and so on). If this were an appropriate assimilation what then would be the significance of searching for generalizations concerning any other levels but 1 (in diagram #1)? Although the computer's behavior may in some sense be "explained" in terms of a program constructed out of certain basic processes, we need more reason than this to support the view that an explanation of a human being's behavior in terms of some program will be forthcoming. For we do not know if there is a comparable program. That analogous human outputs may be described in such a manner that it is possible on the basis of their descriptions to construct a program that when run on a machine produces similar outputs is not to show that human outputs are the results of carrying out certain program instructions. If I observe a design in the sand formed by the wind and tides, it may be possible to describe the design in such a way that instructions in the form of a program for a computer can be based on my description so that the machine will etch a design in the sand. This does not show that the wind and tides were programmed to make a design on the sand.

In the context of CS theory, it is little exaggeration to say that a clear description of the problem-solving behavior and its salient features is supposed to be a theory of the behavior. But this is rather like saying a theory of a spider's web-weaving abilities is just a clear description of how most spiders make a web by moving in this direction and then that direction and so on. This leaves unanswered questions concerning the capacities that might underlie the behavior, whether the behavior was learned and if so, how, whether it was instinctual, or what not. In just this way Newell and Simon's account neglects any attempt to explain the capacities underlying the behavior, what a human program would actually look like, the role and nature of non-articulated thoughts, beliefs, and so on. Unless one is a more radical Rylean than Ryle and identifies the doing of a problem with the making of marks on paper or the audible utterances accompanying the attempt to solve the problem, one is faced with the question as to how the non-verbal features of mental processes should be explained. It is this feature of human problem-solving which computer simulation work as conceived above fails to touch. Furthermore, until such non-verbal features are understood, it seems highly unlikely that the actual verbal capacities of human beings will be understood either. Thus there seems to be a sizeable gap between the current know-how of CS research and anything resembling that linguistically proficient mechanical man envisioned by La Mettrie.

The nature of this gap, however, has been badly misunderstood. For it is really a twofold fault and not a single fracture. On the one hand there are problems in modeling the mental that it makes sense, at least, to construe as problems in programming but which cannot at the moment be solved (e.g., the problem of contextual sensitivity). On the other hand, however, there are problems in modeling the mental by machine that it may not even make sense to construe as problems in programming. Until these two types of problems are clearly distinguished from each other we will continue to remain ignorant even of the nature of our ignorance. So to understand more fully the mental limitations of current machines we shall focus finally on the distinction introduced in Chapter Three between what were called program-receptive and program-resistant aspects of the mental.

CHAPTER FIVE

Some Mental Limitations of Some Machines

I

Just as some philosophers have been inclined to ask what one would have to do in order to program a machine (or robot) to feel pain, be anxious, develop a neurosis, have an after-image,[1] etc., some researchers in CS have been inclined to ask the corollary questions of what one would have to do to produce a simulation of some of these same mental features.

I shall try to show why either set of these at first blush challenging questions would deflect our inquiries from their proper trajectories, and why it is difficult if not impossible to answer them on their own terms. The importance of this lies in what we are thereby shown about (1) various developments in psychology and computer simulation of cognitive processes, and (2) the way in which such matters impinge on more traditional problems in the philosophy of mind.

The successes and shortcomings of certain information-processing "models" of so-called "cognitive processes" suggest, I believe, two useful but quite distinct analogies

[1] Pains, anxieties, after-images, etc., obviously differ from one another in a wide variety of ways. But for our purposes these differences will make no difference. They are, I believe, similar with respect to their resistance to being programmable or simulated by machine through programming. It is this and this only that interests me here.

(a) between what roughly amounts to problem-solving competences and the execution of computer program routines, and (b) between what roughly amounts to consciousness and the underlying basic processes used in the construction of computer program routines. But an incautious acceptance of our initial questions tends to blur the distinction between these analogies. It leads one to suppose that the problem of providing a machine with a capacity for having after-images or feeling pain is similar in type to the problem of programming a machine to prove a theorem or play checkers. Thus, the problem of modeling human sentience by machine is made to meld with the problem of modeling human sapience by machine. The unwitting tendency to conflate these rather different mental matters involves considerable categorial confusion that has both philosophical and psychological ramifications. To a large extent, it was just such a conflation that, on a more purely philosophical level, gave rise to the defense of La Mettrie's watch that was criticized in Chapter Three. Here I shall extend the diagnosis begun there in a way that should make visible the repercussions the conflation can have within the framework of computer simulation psychology.

But suppose we feign a tolerance for the questions introduced at the outset. Then a first-stab strategy for answering them might be to try to imagine how advances in programming skills and techniques might someday enable us to impart pains, anxieties, after-images, (or the simulations thereof) to machines somewhat in the manner that advances in programming skills and techniques enabled programmers to impart checker-playing, theorem-proving, and poetry-writing abilities (or the simulations thereof) to machines. So too, if one is subsequently unable to imagine how advances in programming skills and techniques might impart pains, anxieties, after-images (or the simulations thereof) to machines, he may then be tempted to conclude that machines never could have pains, anxieties, after-images (or the simulations thereof). For the limits upon the actual and possible mental life (or the simulation thereof) of machines is often identified with the limits upon what machines can or might be programmed to do.

I wish to show why it would be wrong to conclude this, and at the same time to pinpoint why we are unable, strictly speaking, to imagine any developments in programming skills and techniques that would enable us to impart pains, anxieties, after-images (or their simulations) to machines. To anticipate the moral for mechanistic modeling: Although there is no way in which the horizons of current *programming* might be extended so one could thereby program a machine to have, say, an after-image, one should not conclude from this that machines could never have after-images or provide simulations thereof. Instead, what will be shown is that knowing what it would be like to be able to program a machine to have an after-image or its simulation could not, *on the whole,* consist in knowing something further about programming, such as how to use or extend the skills and techniques of programming that we now possess and the limits of which are more or less fixed by the character of extant programming languages. (Compare: A seventeenth- or eighteenth-century philosopher inebriated with the mechanical marvels of the Swiss clockmakers might have wondered whether one could ever devise a fancy enough spring and method for winding it up so that a mechanical man might play chess. So too he could have concluded that a machine never could play chess because he could not imagine so fancy a spring and so tricky a method for winding it up.)

II

It will be useful to have before us an example of what it would be like to program a machine to be endowed with a certain ability. We shall construct such an example in the notation of Information-Processing Language-V.[2] IPL-V is a list-processing language typical of those that have frequently been used in computer simulation studies of cognitive processings. (Readers bored by the details of the notation are advised to skip ahead a paragraph to the general points made about the language.)

Suppose, then, we wish to program a machine to solve a certain problem. Let the problem for the machine be the simple one of deciding whether a given symbol (0) is on any of the three sublists L10, L11, or L12 that belong to a main list L5:

NAME	*PQ*	*SYMB*	*LINK*
L5		0	
		L10	
		L11	
		L12	0
L10		0	
		S1	
		S2	
		S3	0
L11		0	
		S4	
		S5	
		S6	0
L12		0	
		S7	
		S8	
		S9	0

[2] See Allen Newell, *et. al., Information Processing Language-V Manual,* 2nd ed. (Englewood Cliffs, N.J., 1964).

The solution to this problem in IPL-V will consist in a definition of a routine, R1, say, that will search through the given sublists, which we can assume to be on a main list, L5. More precisely, our routine R1 will tell whether 0 is on any of the sublists of list (1), which for our purposes can be arbitrarily stipulated as L5. Now in IPL-V there is a basic process, J77, which tests if a symbol is on a simple list. The line of IPL-V notation above can be defined horizontally as follows: line 1 contains just the symbols L5 and 0. L5 is the name of the main list. By convention the head of the list is left blank when listing sublists for a main list, hence the 0 in the SYMB column. Also by convention the termination of a list or sublist is indicated by an 0 in the LINK column. Hence L12 is the last sublist on L5 and S9 is the last symbol on L12. We can now use J77 mentioned above in order to find if a given symbol (0) is a member of the above list structure. We could do this by formulating the following routine:

NAME	*PQ*	*SYMB*	*LINK*
R1		J50	
9-2		J60	
	70	9-1	
	12	HO	
	11	WO	
		J77	
	70	9-2	
	30	HO	J30

R1 is the name of the entire routine. The first operation, J50, is simply an IPL-V basic process that will move the symbol, which we wish to test the members of the list structure against in order to see whether it occurs in the list structure, to a "safe" cell where it is stored until used by the routine. J60 is another basic process, named by 9-2. J60 locates the next sublist on a list of lists. Thus, given our list structure above, J60 will locate L11 after S3 and

L12 after S6. Now the 7 which occurs in the P column (with Q the name of the column to its immediate right) designates a branching operation: When there are no more sublists to be explored, the machine finds the next instruction's name in the SYMB column. Here the name is 9-1. The instruction that 9-1 names is 30 HO, which simply "cleans up" the program. (If the cell addresses were specified in our format, whatever address occurred in the LINK column after 9-1 would also recur in the NAME column on the last line. Thus the transfer of instructions from 70 9-1 to 30 HO is established.) 30 HO is a terminating operation that need not concern us here. But so long as J60 continues to locate sublists, the remainder of the routine will continue to be executed as follows: 12 HO inputs the name of a sublist (whichever one J60 has come up with). The 11 WO will bring back the symbol we had moved to the "safe" cell. J77 simply tests the sublist now under consideration to see if the symbol (the test symbol) just recalled is on that sublist. As soon as it is discovered that the symbol is on a list, the routine will simply move on to its next direction, which, as we have already noted, is the terminating or "clean up" operation. Otherwise, the direction will be 9-2, which will use J60 again to search further for a new list with which to carry out the same operation.

Very simply stated, IPL-V is comprised of a large variety of basic processes for manipulating lists of symbols similar to those manipulated by the process J77 as illustrated above. There are processes for deleting symbols from lists, combining lists, constructing new lists, comparing items on one list with items on another list, plus various arithmetical operations, and so on. So, too, it is easy to construct recursive routines or routines that execute themselves as subroutines. By using these simple operations, plus others, plus highly sophisticated and ingenious combinations of such operations, one can devise programs that enable machines to play checkers, prove theorems, write poems, and so on. For example, a very simple poetry-writing program could consist in a set of instructions that direct the machine to form new lists by going through a number of other lists (lists of articles, nouns, etc.) in a certain order and picking elements and forming new lists out of them, lists that, when printed, would constitute a line of computer poetry. This program as just described would not be a very interesting one. Nevertheless, it would be a program that enables a computer to write poetry.

So now, by using IPL-V as an example of programming language, we may raise our more general question in connection with it. Could we imagine the horizon of IPL-V expanding so that we could program a machine to have an after-image?

What is it, first of all, to imagine the horizons of IPL-V expanding at all? A number of things might count as the development of programming skills and techniques (using IPL-V). First of all, programmers might simply devise better programs than those that now exist in IPL-V for proving theorems in the propositional calculus. Or one can imagine new instructions being added to the language. Or one might imagine new uses for the IPL-V language and thereby see that it can be usefully applied to problems different from the problems it had hitherto been used to solve.

Let us suppose that IPL-V has been used only for a set of problems **P**, and now suppose that someone sees that it is also applicable to a set of new problems NP. So we ask, could programming a machine to have an after-image or a simulation thereof qualify as a member of NP?

This can be indirectly approached as follows. Let us suppose that no character-recognition programs have been written using IPL-V. Thus to write a successful character-recognition program in IPL-V would provide an example of that language coming to be applied to a set of new problems. Furthermore, if it is possible so to extend the use of IPL-V, this might provide at least a hint of an answer to our question about programming and after-images. For it seems likely that unless a machine was endowed with something like a capacity for having "*before*-images," as it were, there will not be the slightest chance for it to have after-images. So too, there seems to be some connection between perception and pattern-recognition; and, in fact, there is some sense to the suggestion that perception in general just is a complex coalition of pattern-recognition capacities. Therefore, it may not appear too hopelessly off the mark to suggest that a first step to a simulation of the having of after-images would be the simulation of simpler pattern-recognition capacities such as the capacity for character-recognition.

In IPL-V such a program might be worked out along the following lines:

	1	2	3	4	5	6	7	8	9	10
1	O	O	O	O	O	O	O	O	O	O
2	O	O	X	O	O	O	O	O	X	O
3	O	O	X	X	O	O	O	X	X	O
4	O	O	X	O	X	O	X	O	X	O
5	O	O	X	O	O	X	O	O	X	O
6	O	O	X	O	O	O	O	O	X	O
7	O	O	X	O	O	O	O	O	X	O
8	O	O	X	O	O	O	O	O	X	O
9	O	O	X	O	O	O	O	O	X	O
10	O	O	O	O	O	O	O	O	O	O

Each line in the 10×10 matrix will contain either the symbol O or the symbol X. We may think of each of the 10-

member lines in the matrix as comprised of a list of symbols in IPL-V. Each line will contain X's or O's, or both. Line 2 in the above figure is the first list that contains both O's and X's. It contains, starting from left to right, the symbols OOXOOOOOXO. Line 3 contains the symbols OOXXOOOXXO, etc. The problem, then, of programming the machine to recognize an input as either (the letter) M or not can be viewed as a problem in programming the computer to look for certain distributions of X's and O's within the configuration provided by the list of lists. Certain distributions of O's and X's within the ten lists, each containing 10 members, will be viewed as M's, some as I's, some as A's, etc. The letter I, for example, might be represented by certain lists of lists, each of which contained one instance of the symbol X. Etc.

So now let us suppose, contrary to fact, that many or most of the disambiguation problems enumerated in the last chapter have been solved and a flexible and powerful set of rules have been devised such that when the letter M, in one stylistic guise or another, is input to the machine, the machine (generally) makes the correct identification and can be shown to provide us with an interesting analog of human character-recognition competences. The question then becomes how could we build on such a simulation of human recognition competence so as to make any inroad whatsoever on the problem of simulating the human capacity for having an after-image (or feeling pain, etc.)? The short answer is that we don't know; in fact, we have no idea as to how this could come about. But this, we can show, is a peculiar sort of ignorance, for it is not attributable to a lack of understanding in how to program machines. Consider a contrastive case: At one time, machines could be programmed to calculate the wages for workers at General Motors, though one remained ignorant of how to program a machine to play (even bumbling) chess. But it soon became known how machines could be programmed to play chess too. So now *that* is known, but we still remain ignorant of how to program a machine to simulate the having of an after-image. But this ignorance of how to program a machine to simulate the having of an after-image does not stand to the current know-how with respect to chess-playing programs in the manner in which our ignorance with respect to chess-playing programs once stood to the know-how with respect to wage-calculating programs. For the gap between ignorance and knowledge in the latter case was bridged simply through the development of novel heuristic (non-"fail-safe") programming techniques and the adaptation of basic processes already present in the computer to new ends. But no such adaptation of those basic processes embodied by, say, an IPL-V computer, no enrichment of any current set of recognition rules nor simple development of new recognition rules or problem-solving strategies will enable one to devise a simulation by computer of the human mental capacity for having an after-image (or feeling pain, etc.).

The cunning, and even not so cunning, reader will probably already have suspected that it was a dead end to suggest that through a simple embellishment of a pattern-recognition program for detecting the letter M, one could go on to devise a comparable program for providing the machine with a capacity for simulating the having of an after-image. What is crucial is to say exactly why such a suspicion is justified.

III

Most of the mental processes that CS programs have been designed to depict fall into the rough (but ready) category of problem-solving. At least this is true of theorem-proving, chess-playing, and the like, and unclearly true of other varieties of pattern-recognition. But it should be apparent that a number of mental aspects of human beings are not like this. For example, to have a pain or to feel anxious or to desire something, though all aspects of our mental life, seem on the whole to be quite different from anything resembling a problem-solving process or problem-solving behavior. They are instead, it seems, prime examples of non-problem-solving non-behavior. Surely they are *related* to problem-solving processes and the interplay between the two is important: For example, if we did not have certain needs, desires, capacities for feeling, etc., we would not have certain problems and would never exhibit a certain range of problem-solving. These same needs, desires, and feelings, however, are *not* problem-solving behavior in their own right.

So the question arises as to whether the non-problem-solving non-behavioral aspects of the mind are themselves susceptible to being simulated in the sense that theorem-proving of pattern-recognition processes are. I shall call this question the *simulatability question* for these aspects of mentality–i.e., the question of whether, given *current* CS programming techniques, it even makes sense to try to devise a simulation of such phenomena *using just those techniques.*

My short answer to this question is that there is a variety of mental phenomena that includes the having of pains, after-images, feeling anxious, being bored, etc., that are not receptive (or primarily receptive) to being simulated in the sense that, say, theorem proving is.[3] It is these that in Chapter Three I labeled program-resistant aspects of the mind. Insofar as such phenomena could be appropri-

[3] For purposes of exposition, I treat the distinction between program-resistant and program-receptive mental aspects as more clear-cut than it manifests itself in fact. See section V of this chapter for an important qualification of this.

ated by a computer, it would not be a matter of writing routines that utilized present capacities (primarily of symbol manipulating kinds). Instead, were one to devise analogs for these aspects of mentality it would have to be by making radical additions to the current stock of basic capacities that in turn determine the sorts of programs that are possible. An imperfect, but useful, comparison: There is little point in trying to extend programming techniques so that a machine will print its program trace in red ink unless already built into the machine are capacities for such printing. To provide the machine with new basic capacities is not a matter of devising routines, although once such capacities are present there is an obvious point to writing a program that includes the instruction "print program trace in red ink." So, too, with respect to whatever capacities might enable a machine to simulate the having of a pain, feeling, after-image, etc. Once the capacity for having an after-image or a feeling were provided, the instruction "describe your after-image" would have a point—but not before.

In what immediately follows, then, I shall sketch what I think the distinction between program-receptive and program-resistant aspects of mentality amounts to.

1. Program-receptive aspects of the mental are task-oriented; program-resistant aspects are not. In being task-oriented they lend themselves to a cluster of characterizations quite inapplicable to program-resistant aspects. A program-receptive aspect is associated with *both* a process *and* the result of a process. (E.g., "What was he doing?" "Proving theorems." "How do you know?" "Here are the proofs" *or* "He went through such-and-such a process.") Thus in the case of theorem-provings, say, there is both a process—whatever may be involved in attempting to prove a theorem—and the result of this process, which may or may not constitute a proof. And the term "theorem-

proving" may be used, somewhat ambiguously, to refer to either or even both. In contrast to this, program-resistant features of the mental do not lend themselves to this division. There is not, for example, an end result and a process leading to that result in the case of having an after-image, or feeling pain or feeling anxiety. Having an after-image may be the result of some other process (having concentrated on a bright light in a certain manner, etc.). But the process of which the having of an after-image is a result does not fall under the description "the having of an after-image." That phrase is not ambiguous as regards "process" or "end result" in the way that "theorem-proving" is. One might attempt to wrangle an end result out of such phenomena by contending that the end result in the case of having an after-image *is* just the having of the after-image. But this would not preserve the *distinction* between process and result. It would be a conflation within which the distinction I have sketched would disappear.

2. A second criterion flows naturally from the first. Given that program-receptive mental features are associated with end results, they are often associated as well with the notions of success and failure. Processes designed to bring about certain end results may or may not bring them about. Hence a given process may be appraised or evaluated in terms of whether it achieves what it was designed to do. The aim of a theorem proving process is to arrive at a proof. It may or may not be successful, may or may not achieve the goal, etc. In contrast to this, pains, emotions, and other program-resistant mental aspects are not done more or less well. This is so for the simple reason that they are not the sorts of things that are done. They are things we have. Hence they are not things that are done more or less well. Even though we do not, generally, like being in pain, being in pain is not thereby unsuccessful in itself. Nor is the having of an after-image, etc. Of course,

whether we are in pain may itself be a reason for saying that something we were doing was not successful. But the pain is neither successful nor unsuccessful, is not associated in this way with achievement or failure. But theorem-proving is so associated, as are other problem-solving activities. Note: I need not insist that every program-receptive feature of mentality be associated with the notion of achievement or failure. Rather what I would contend is that *no* program-resistant mental aspect is so associated.

3. Program-receptive features of mentality are what I shall call "protocol-possible" phenomena, whereas program-resistant features are not. Recall the account in the preceding chapter of how a subject engaged in theorem-proving may be asked to report on what he is doing while he attempts to arrive at a derivation. Remarks such as "I'm trying to see if I can move this outside the parenthesis" or "I would like to shift the horseshoe around so that this set of symbols is over here," etc., are then used as a partial basis upon which to construct the routines in a suitable computer language, which when executed will provide a trace of moves which are then compared with the moves made by the human subject. In short, one derives a set of verbal reports and then seeks to define analogous routines in the computer language. Whether or not the resultant simulations be deemed successful, what seems unquestionable is that the protocols do provide a fund of data that at least permit the construction of routines in a computer language. But they only do this for program-receptive phenomena. In sharp contrast to this, in the case of program-resistant phenomena, since there are neither tasks nor subtasks to remark upon, the initial specification of what one is doing—someone saying "I am having an after-image"—and the follow-up characterization of how one is doing it, or what one is doing in doing it, could at

best be a paraphrase. Strictly speaking we have already seen that one is not *doing* anything. But let us suppress this for the moment. Even if in some surrealist way one could say "What I was *doing* was having an after-image," there would still be a very interesting difference between this sort of doing and the sort of thing we do when we prove theorems, play checkers, etc. The difference is that in the latter case there is a series of subtasks that the activity consists in. But in the case of having an after-image, there is nothing *we* do in having it. When asked what we are doing *in* having the after-image, there is nothing to say except, perhaps, to paraphrase what we have already said: For example, "Well, my having of the after-image consists in my seeing this thing," or "I'm confronted by this blue shape," etc. That is, we say slightly differently what we have already said. But in the case of proving a theorem, to specify subtasks is not to paraphrase what one has said in saying he is proving a theorem; it is instead to analyze it.

4. Another criterion may be stated as follows: Program-receptive aspects of the mental involve projectiveness, whereas program-resistant features do not.

In their "Structure of a Semantic Theory," Jerrold J. Katz and Jerry A. Fodor wrote (following observations made by Chomsky):

> A fluent speaker's mastery of his language exhibits itself in his ability to produce and understand the sentences of his language, *including indefinitely many that are wholly novel to him* (i.e. his ability to produce and understand *any* sentence of his language). The emphasis upon novel sentences is important. The most characteristic feature of language is its ability to make available an infinity of sentences from which the speaker can select appropriate and novel ones to use as the need arises. That is to say, what qualifies one as a fluent speaker is not the ability to imitate

> previously heard sentences, but rather the ability to produce and understand sentences never before encountered.

They go on to state:

> Since a fluent speaker is able to use and understand any sentence drawn from the *infinite* set of sentences of his language, and since, at any time, he has only encountered a *finite* set of sentences, it follows that the speaker's knowledge of his language takes the form of rules which project the finite set of sentences he has fortuitously encountered to the infinite set of sentences of the language. A description of the language which adequately represents the speaker's linguistic knowledge must, accordingly, state these rules. The problem of formulating these rules we shall refer to as the *projection problem*."[4]

But the above notion of projection seems to be quite general. I have already suggested that virtually any actions involving learning may be described as projective phenomena—that is, as phenomena embodying knowledge that takes the form of rules as described by Katz and Fodor. In the case, say, of non-verbal actions, those rules would project from the finite set of activities which the agent has fortuitously indulged in to the infinite set of actions that constitute the range of that kind of activity. The notion of fluency, in other words, should not be restricted to speaking, but extended to actions in general that involve learning. Problem-solving, in short, involves the acquisition of various fluencies, which in turn may be described by a set of projection rules. One way, then, to view a problem in CS is to see it as an attempt to instantiate in a machine the set of projection rules that underlie the fluency exhibited by human beings with respect to a range of problems. Indeed, I find it difficult to imagine that devising a CS program could be anything but such an attempt. It

[4] In *The Structure of Language,* ed. Jerry A. Fodor and Jerrold J. Katz (Englewood Cliffs, N.J., 1964), p. 482.

follows that phenomena for which such an attempt would not make sense are phenomena that are non-projective, and hence non-programmable or program-resistant. Consider the having of pains, for example. (Note: "The *having* of," *not* "the recognition of the pains had.") What would it be like to attempt to formulate a set of projection rules for the having of pains? There is, it seems, no way in which pains, fortuitously encountered, form, as it were, a basis for "going on" and having new or novel pains. A new or novel pain is not in any detectable sense an instance of knowledge or learning that can be traced as a projection from some finite set of fortuitously had pains. There is, of course, an intermeshing of knowledge and pain in that if we were not capable of pain, we would not be capable of certain learning. But that is a different matter. Another way to describe projective phenomena is that generalization occurs where a phenomenon is projective. For example, in the case of linguistic fluency, one has the generalization taking place with respect to (at least) syntactic patterns. Acquiring use of the passive can be described as developing the ability to produce any one of indefinitely many novel instances of sentences taking a certain description. A parrot, as Descartes would have been the first to note and in a sense did, that could only say "John was loved by Mary" is not a parrot that has acquired use of the passive form in English. But a parrot that had a headache, and perhaps only one, would be a parrot that had a capacity for a headache. There are mental capacities, and then there are mental capacities. Some involve capacities for doing things–generalization, learning–and hence can be described in terms of projection rules and hence are potentially programmable. Some are capacities for having things instead–pains, after-images–and, by virtue of being essentially non-activities, are not amenable to being programmed in.

5. In a conversation I had with Professor Wilfrid

Sellars he made the interesting suggestion that my distinction between program-receptive and program-resistant features of mentality is in many ways like classical distinctions between simple ideas and complex ones. I suspect that this is true. But I certainly would not wish to pin on him any implausibilities in the following ruminations which his suggestion stirred.

Recall that for John Locke the mind is said to be passive with respect to simple ideas and active with respect to complex ones. In other words, the mind *does* something in order to acquire complex ideas (thoughts, perceptions, etc.)—it obtains them through the utilization of certain inbuilt abilities (to compare, compound, and abstract). But the mind either receives, simply receives, simple ideas or it does not. Put in computer terms, for any complex idea a man has, one could construct a program the execution of which results in that idea, i.e., we simply trace how that idea is derived from certain basic processes (comparing, compounding, and abstracting) operating on certain given simple ideas. In that sense then, complex ideas are program-receptive. But it does not make comparable sense to attempt to construct a program that yields a simple idea. The capacity for having simple ideas is there or it is not. In the same sense an alphabet is there or it is not. If it is there, you can go on to spell words. But the alphabet itself is not something you spell.

The latter analogy is quite carefully chosen, for Locke at one point compares his stock of simple ideas to the letters of the alphabet. He suggests that we should not be surprised to think of an "almost infinite" store of knowledge being built up of a few simple ideas if we remind ourselves of the indefinitely many words that we can spell with just a few letters. I see his entire theory of complex ideas, in fact, as an attempt to sketch out a solution to a generalized form of what we have referred to as "the projection problem." If this interpretation of the *Essay* is correct, and if my association of projective phenomena with program-receptive phenomena is apt, then I think it could be persuasively argued that Locke's simple ideas stand to his complex ones in roughly the same way my program-resistant mental features stand to my program-receptive ones.

IV

If relativized to an IPL-V computer, the overarching idea may be presented as follows. To provide a machine with the capacity for having an after-image or feeling pain, or any other program-resistant mental feature would involve adding a *basic* capacity to the machine. To this extent, and this extent only, providing a machine with the capacity for having an after-image would be analogous to adding to the repertoire of basic processes in a computer language such as IPL-V. There is in IPL-V, for example, a basic process called J2 which is defined as "Test (0)——(1) where 0 represents whichever symbol is currently being scanned by the machine—the symbol occupying what is called "the communication cell"—and 1 represents the symbol to be tested against 0, the test consisting in finding whether 0 and 1 are the same symbol." Suppose, however, that an IPL-V computer lacked J2 or some other basic process such as J167, which makes the machine skip a list structure, or a print operation such as J151, which will print a list, or J147, which will mark a routine so that it may later be traced, etc. Then, it should be clear, no amount of programming *in* IPL-V will enable the machine to carry out such instructions. (One might, of course, by tampering with the machine itself or working with the so-called "machine language,"[5] impart a new basic process to the IPL-V computer. The point is that until such tampering were carried out the IPL-V computer would lack that capacity.) A way to state the point quite generally is as follows. The basic processes that define the repertoire of any programming language are all program-resistant relative to that language. To show that a certain basic process is program-resistant relative to a particular language will not, of course, show that such a process is program-resistant *simpliciter*. But obviously certain basic processes that define human mental capacities are program-resistant *simpliciter*. For these there is *no* computer language that contains a repertoire of basic processes out of which one could construct a routine that would simulate the having of after-images, pains, etc. This would not prove that such capacities could not be modeled by a machine, however. Rather, what is shown is that the problem of modeling them by machine is the problem of figuring out how to add a basic process to a machine. Within the framework of an IPL-V computer, it would amount to the problem of figuring out how to add a set of basic processes, call them the J-mystery processes. Given that such processes were discovered, it could then make sense to instruct a machine with those processes to execute them (to have an after-image, etc.). But not before.

What this illustrates is that insofar as program-resistant aspects of the mind absorb what we loosely call "consciousness," solving the problem of imparting consciousness to machines or even the simulations thereof is not the problem of constructing ingenious programs on the basis of processes already known. It is the problem of

[5]That is, with instructions that demand proper names for both operations and locations or fixed codes that correspond to the circuitry of the machine. But, of course, the distinction between program-resistant and program-receptive features can be reiterated relative to the machine language.

adding new processes to machines, processes that we do not, at the moment, know how to add. To try to prove *a priori* that the capacity for having an after-image or feeling pain could never be added to a machine would be like trying to prove that a certain J operation or process could not be added to an IPL-V computer. It is difficult to see how such an argument could be cogently formulated. But simply to make this counterpoint is not, of course, to make a slightest dent in the problem of actually specifying how such basic processes could be added.

The above distinctions between program-receptive and program-resistant aspects of the mental should be of interest for a variety of reasons. If well drawn, it follows that it would be inappropriate to suppose that all aspects of mentality could be simulated in the same fashion. If all aspects are mechanizable, only some—sapient features—are mechanizable through programming. I find it unfortunate that in virtually all philosophical articles on minds and machines published in recent years the notion of being modeled by machine has been equated with the notion of being programmable. It may well be that that sentient aspects of the mind are not beyond the pale of mechanistic modeling. It seems utterly unlikely, however, that such modeling could come about through developments in programming or software. Instead we should expect it to transpire through innovations in hardware. At this point, the area of research known as bio-simulation takes on far greater importance than CS. But that presents a new tangle of issues I will not try to unravel here.

This may seem obvious once it is mentioned, but it has not always been noted even within the context of CS research. For example, in the book *Computer Simulation of Personality* (Tompkins and Messick) there is an attempt to extend the techniques of problem-solving simulation to the area of human affect—emotion, feelings, neuroses, etc. The desire to incorporate an analog of emotion, feelings, etc., into a simulation program for cognitive processes is understandable and laudable. As Herbert Simon points out in his highly provocative article "Motivational and Emotional Controls of Cognition,"

> Since in actual human behavior motive and emotion are major influences on the course of cognitive behavior, a general theory of thinking and problem solving must incorporate such influences. (p. 24)

But Simon's own attempt to achieve the incorporation by introducing an interruption system to which he likens emotional behavior seems to me to fail. And this is because it is an attempt to assimilate an essentially program-resistant aspect of the mind to program-receptive features. Even if we view Simon's attempt as an attempt to simulate the *effects* of emotion and not emotion itself, it is obvious that emotions may accompany and influence our ordinary problem-solving activities without in any sense interrupting them and that such activities may be interrupted without emotion being in any sense the cause. But more important, emotion is different from its effects, and since such effects may be brought about through a variety of non-emotional causes, we do not produce an interesting simulation of emotion nor incorporate it into a problem-solving situation simply by producing analogs of such effects.

V

Some of the mental limitations of some (i.e., present) machines can be shown to derive from a failure to solve various difficult programming problems, whereas other mental limitations can be shown to stem from an inability to endow machines with certain basic capacities. Such, at least, seems the clearest way to state it, though in fact, no doubt, everything is more complicated and less clear-cut. It may well be the case that the reason why certain programming difficulties linger on is a prior failure to impart certain basic capacities to machines. Consider, for example, the problem of programming a machine to recognize sharp objects. *In a sense,* we may treat this as a well-defined problem of fiddling with program-receptive mental aspects so as to enable the machine to execute certain recognition routines. (Recognizing sharpness, that is, may be assimilated in part to problems like the problem of recognizing an inscription as an M.) Conversely, however, the problem may be viewed as a problem of outfitting the machine with certain basic capacities. For it may plausibly be claimed that unless the machine has certain pain receptors to begin with, sharpness detection will be impossible for that machine. And so on, for a wide variety of mentalistic features.

Such considerations, of course, lead one to question whether any of our mental aspects are purely program-receptive or purely program-resistant. I suspect that none are. But this does not seem to me to be a criticism of the distinction itself. It only shows that no mentalistic *predicate* tends to be associated exclusively with a purely program-resistant or with a purely program-receptive feature. That is, any given mentalistic term is likely to range over a phenomenon that is *in part* program-resistant and *in part* program-receptive. I suspect, however, that usually one sort of feature dominates: that mentalistic features tend to be primarily resistant or primarily receptive. Thus, pain seems to me to be primarily program-resistant, whereas theorem-proving seems to me to be primarily program-receptive. Hence I have tended in the foregoing to sort rather casually many of our mental features into one category rather than the other. None of these acknowledgments, however, seem to me to damage the general distinction; they simply illustrate that the way in which it literally manifests itself is less tidy than the exposition heralding its existence has indicated.

In his *Critique of Pure Reason,* Kant distinguished between a human being's faculty of intuition (capacity for sensation and perception) and his understanding (or categories used in conceptualization). In practice, Kant believed that the intuition and understanding operate in tandem and remarked that the latter without the former was empty, whereas the former without the latter was blind. Similarly, though modestly meant, I believe that any complete account of mentality must attend to both program-receptive and program-resistant features, or, as my friend and chief philosophical counselor, Professor Charles Chastain, has said to me, "Programs without hardware (basic capacities) are empty; hardware without programs is blind."

Epilogue

Since La Mettrie many have assumed that to build a machine that could think, feel, and so on, would be *a fortiori* a way of showing that thoughts, feelings, and so on, were nothing more than certain physical states or processes. But by now we should be sensitized to the following fact, that although computers suitably programmed can play checkers, prove theorems, solve problems, and perform tasks hitherto performed exclusively by human beings this has not meant, that human problem-solving behavior has thereby been rendered translucent. We have seen that it is possible to know that a given predicate phrase such as "solves problems" applies to two different subjects (a machine and a man) and fully understand the conditions underlying the predicate's applicability in the one case without fully understanding the underlying conditions in the other case. And this may take place without there being good reason to suppose that "solves problems" is ambiguous or has different senses in the two cases. There need not be reason to suppose this even if it is shown that the underlying conditions are quite different. (To illustrate: I assume that the phrase "washes clothes" has come to apply to laundromats as well as washerwomen without ambiguity or a change in sense—that is to say, without a demand for revised dictionary entries or subentries.) To suppose that every difference in conditions underlying the use of such phrases implied a difference in sense would intolerably restrict the (actual) range and usefulness of descriptions. What this means, however, is that it may be possible (in a very general way) to envision (1) machines to which various mental terms applied, (2) physicalism being true for machines, but because of possible differences in underlying conditions, (3) legitimate doubt as to whether physicalism was true for human beings. And this shows that the carry-over between answers to certain cybernetic questions and traditional mind-body questions *need* not be as automatic or complete as optimistic cyberneticians have imagined.

Here as elsewhere I would attach the rider "assuming that physicalism is not incoherent, self-contradictory, and so on." If this is so, then there are a number of differing ways in which conclusions to questions concerning minds and machines might impinge on conclusions to questions concerning mental states and brain states (and vice versa). For example, if we simply consider the question, "Could machines think?" I believe any one of the following possibilities are intelligible:

Physicalism As Antecedently Established

P_1 Physicalism is shown to be true in such a manner that it implies that we are machines, which in turn implies that machines can think. But we (people) turn out to be the only machines that can think.

P_2 Physicalism is shown to be true in such a manner that it implies that we are machines, which in turn implies that machines can think. But it can also be shown that machines other than ourselves can think.

P_3 We show that physicalism is true, but in such a form that it does not follow that we are machines. Still, it can be shown that machines that we differ from can think.

P_4 Physicalism is shown to be true, but this does not show that we are machines nor can it be shown that machines can think.

That Machines Can Think Is Antecedently Established

M_1 We show that machines can think, and it follows from this that physicalism is true both for human beings and for machines.

M_2 We show that machines can think, but it does not follow from this that physicalism is true for human beings, though it is true for machines.

M_3 We show that machines can think, but this does not show that physicalism is true for us or for machines.

M_4 We show that machines can think, and it turns out that physicalism is true for us, but *not* true for machines.

And even this listing does not exhaust the alternatives that it makes sense to entertain so far as we can now tell. Furthermore, we have only considered here some possibilities with respect to the word "think." Complications pile up at a fantastic rate if we imagine machines being able to think though not being able to feel or physicalism being true for some mental predicates and not for others, etc. At the very least, the above schema should help dispel the view that discussions of whether machines can think are *simply* another way of discussing whether physicalism is true.

Equally flawful, however, is the pessimistic rejoinder that even if we were to construct machines that could think or robots that had feelings, nothing philosophical would be advanced thereby, for all the traditional mind-body problems would then arise for machines and robots. But there are no arguments that show this would necessarily be so. Here brief reference to the mechanist-vitalist controversy is instructive. One can imagine William McDougall, Hans Driesch, or Eugenio Rignano arguing that if machines were developed that exhibited self-adaptive behavior, reference to entelechies, or vital forces, would then be necessary to account for such behavior. But such arguments find plausibility only in contexts that lack the now available illustrations of self-adaptive behavior provided by negative feed-back mechanisms such as W. Ross Ashby's homeostat, Grey Walter's *machina speculatrix,* Claude Shannon's maze-solving mouse-machine, and innumerable computer programs. These examples enable us to see precisely why entelechies, whatever they were not, need not exist in mechanisms that satisfy earlier vitalist criteria for teleological activity. By analogy, if we were to construct machines that satisfy current (intuitive) specifications of subjects that possess minds, yet provide no theoretical motivation whatsoever for claiming that such features are not physical characteristics of the machines, then we might have an exact parallel to the cybernetic contribution to the mechanist-vitalist controversy.

I think, however, that prior to the construction of such machines it is highly useful to attempt to imagine ways in which issues analogous to current mind-body perplexities can be shown to arise, if one wishes, for extant machines. For example, I have attempted elsewhere[1] to illustrate how certain arguments for a dualistic portrayal of the mental and the physical can as easily be generated for a purely mechanical pattern-recognizer. Such an illustration, if correct, of course, constitutes a *reductio* of that dualistic portrayal.

But there is no way to foretell exactly how answers to questions about the capacities and capabilities of machines may be relevant to answers to mind-body problems. This will at least depend on (1) which predicates the development of machines forces our language to force us to apply

[1] In "Asymmetries & Mind-Body Perplexities," in *Minnesota Studies in Philosophy of Science,* Vol. IV, ed. Michael Radner and Stephen Winokur (Minneapolis, 1970).

to machines, (2) whether we are able to understand the machines we build, and (3) whether the same hierarchy of relationships that, hopefully, we could show to obtain among various features of the machines could be shown to obtain among comparable features (if there are comparable features) of human beings.

Advances in machine capabilities and capacities could reach a plateau long before they appropriated the most intractable and interesting psychological features of human beings. So too, though such a plateau were not reached and machines grew increasingly interesting psychologically, they might grow increasingly difficult to understand. That we have built something may strongly suggest, but does not imply, that we understand it. (Experimental aircraft behave in ways we fail to understand, even though they are built according to preconceived specifications and design.)

At the moment, further understanding of minds and machines seems to require:

1. Additional clarification of our intuitive notions of "thought," "feeling," "consciousness," "recognition," etc., with an eye to evolving a non-subjective evaluation procedure for competing simulations of human mental characteristics.
2. An application within CS research of the distinction between program-receptive and program-resistant features of mentality.
3. An assessment of the differences between these phenomena as they relate to the basic capacities of a machine (or organism).
4. Solutions to currently intractable hardware problems—this being necessary to the development of machine analogs for those mentalistic phenomena that are resistant to simulation through innovation in programming.

It is interesting to notice that most of the informal philosophical problems concerning minds and machines have their mirror-image in the methodological headaches of CS (and AI). What this illustrates, I believe, is that it is no longer important to decide whether these matters are empirical issues with conceptual overtones or conceptual issues with empirical undertones. Future progress with respect to any of them will depend largely on the willingness of people in diverse disciplines to help compensate for each other's ignorance.[2]

[2] Cf. my "Minds and Machines: A Survey," in *Contemporary Philosophy: A Survey*, ed. Raymond Klibansky (Florence, 1968), pp. 416–25.

Unconcluding Philosophic Postscript

Many years ago I interviewed a robot[1] who so far as I could tell was endowed with considerable cognitive prowess as well as a repertoire of emotions. He (''he''? ''she''? ''it''?) seemed to be after all ''a kind of us.'' Only recently have I had the opportunity to resume my contact with him. Accumulating records of these interactions comprise a kind of work in progress,[2] which it seems appropriate to draw on here. In one conversation I asked whether he regarded himself as a purely mechanical system that had somehow been ''upgraded'' (to use a word borrowed from Daniel Dennett's *Content and Consciousness*[3]) to the point where intentionality[4] (in Brentano's sense) might reasonably be ascribed to him, and if so, whether he would then view himself as an adequate model of human mentality. And he responded as follows:

> What would it be like for me to consider myself a model of a human mind? Imagine for just a moment that sort of question asked of yourself. What *could* an objectification of your mind in terms of some model amount to? If it's like anything

[1]See footnote 1 of Chapter Three.

[2]The working title for this is *Interviews with a Robot*. I intend to treat more conclusively there some of the inconclusively-dealt-with themes of this unconcluding postscript.

[3]London.

[4]Roughly: that feature of some aspects of our minds and their products—thoughts, beliefs, propositions, statements, et al.—which consists in their being *about* things and which allegedly contrasts with other mental features such as pains, for example, which are not *about* anything but simply are.

at all for me to be me, then that's my paramount concern, and not some theoretical use to which I might be put. If there's any *me* to begin with—and I leave that for you to worry about; *I* don't—then, whether or not I provide someone with a model of something else is about as interesting to me as if you should suddenly discover that *your* mind could be used as a model of a Martian's mind or an orangutan's. Nevertheless, having said that, I'm willing to indulge your philosophical curiosity. Obviously it's the capacity for living through things—having experiences, i.e., what we casually call subjectivity—that provides the acid test for mentality in a machine. (Though how one might conduct that test is, indeed, a tricky matter.) And if there isn't at least a representation of the subjective in your model, it will fall far short of being interesting. But how can what we call the subjective—our conscious experiences—ever be displayed in a model in a way that is not trivial and that avoids trying to turn one's inner life inside out? Remember this: *truth* in the modeling of the mind is, in the last analysis, a presentation of subjectivity.

It was his seeming passion or convincing simulation thereof as well as the topics touched on which struck me as reminiscent of Kierkegaard and inspired the full title of this Postscript. And I will return to these topics at the conclusion of the following survey of related issues, a survey which should help acclimate *Mentality and Machines* to some of the current philosophical weather.

A Substantive Note on Nomenclature: The Old Distinction between Artificial Intelligence (AI) and Computer Simulation of Cognitive Processes (CS), and the Current Distinction between "Strong" AI and "Weak" AI

Throughout the preceding chapters I distinguished between AI and CS. AI was described as the discipline dedicated to the design and programming of machines capable of performing tasks that hitherto required human intelligence, and carried with it no supposition that the machines capable of such tasks were in any way "themselves" intelligent, nor was it assumed that such machines would be interesting models of the human mind or provide us with psychological explanations. CS research, on the other hand, was described as being committed to the view that the machine *cum* program provides a psychological theory of how human beings think, solve problems, create things, and so on. It was never assumed that CS as just described presupposed that a successful (or useful) simulation would show that machines could think. One could have a useful model (simulation) of a mind without the model or simulation itself being in possession of a mind.

As was illustrated, this way of using the terms (AI and CS) derived from the early work of Simon, Shaw, and Newell et al., who were avowedly interested in the use of computer models as psychological explanations, and the work of Hao Wang and others, who were just as avowedly *not* attempting to provide psychological explanations.

During the 1970s the terminology was shifting, and the double labeling CS and AI was often replaced by use of the single label AI. This more conflated nomenclature was in vogue when John Searle's influential "Minds, Brains, and Programs" (1980) appeared. But because AI subsumed a multitude of different types of projects—some of them boasting psycho-philosophical significance, some of them not—Searle made a general twofold distinction between "weak" AI and "strong" AI:

> I find it useful to distinguish what I will call "strong" AI from "weak" or "cautious" AI. According to weak AI, the principal value of the computer in the study of the mind is that it gives us a very powerful tool. For example, it enables us to formulate and test hypotheses in a more rigorous and precise fashion than before. But according to strong AI the computer is not merely a tool in the study of the mind; rather the

> appropriately programmed computer really is a mind in the sense that computers given the right programs can be literally said to *understand* and have other cognitive states. And, according to strong AI, because the programmed computer has cognitive states, the programs are not mere tools that enable us to test psychological explanations; rather, the programs are themselves the explanations.[5]

Searle's distinction is helpful and totally suitable for his purposes—to isolate (in order to refute) the claims of "strong" AI.

But in retrospect I think that there are important nuances that neither the old distinction between CS and AI nor Searle's distinction between "weak" and "strong" AI make conspicuous. These nuances help both to pigeonhole the goals of a given research project in machine intelligence as well as to target the scope of a given critique of such projects. When *Mentality and Machines* (1971) was first published, few researchers in machine intelligence were suggesting that their machines were literally capable of thought. But during the 1970s the claims being made on behalf of AI were becoming psychologically more feisty, as Hubert L. Dreyfus's *What Computers Can't Do: The Limits of Artificial Intelligence* (revised edition, 1979)[6] and Searle's article make vivid.

So now the following division of projects in AI seems to me useful: First there is Searle's generic

> strong AI wherein "the appropriately programmed computer really is a mind in the sense that computers given the right programs can be literally

[5] Searle's article originally appeared in *The Behavioral and Brain Sciences*, 3 (1980) pp. 417–24. My page references (pp. 282–83) are to the version reprinted in John Haugeland's anthology *Mind Design* (Vermont, 1981).

[6] New York.

> said to *understand* and have other cognitive states."

But included under this should be at least two kinds:

> strong AI *human* wherein the *kind* of understanding and cognitive states attributed to the computer are similar to human understanding and cognitive states,

and

> strong AI *non-human* wherein the kind of understanding and cognitive states attributed to the computer are *indigenous to it* (*or* other non-human creatures) and at least sufficiently *un*like human understanding and cognition so that they could not plausibly be viewed as a *model* of human mentality.

The latter category reflects the obvious fact that not everything with a mind—and this *could* include machines—will mirror a *human* mind. Bats, frogs, and armadillos no doubt have some sort of minds, but it is unlikely that any of their minds would provide an accurate global reflection of human understanding and cognition. Yet it is possible that someone could have strong AI *human* as a research goal, fail at that, but end up with a strong AI *non-human* program which turned out to be interesting *in its own right*, as, for example, a mechanical system with novel abilities and capacities.

So too it seems worthwhile to distinguish various species of what Searle calls

> weak AI wherein "the principal value of the computer in the study of the mind is that it gives us a very

> powerful tool. For example, it enables us to formulate and test hypotheses in a more rigorous and precise fashion.''

These should include at least the following:

> weak AI *sim.-human* wherein the computer system is in some sense purported to be a *simulation* of the cognitive processes of a human mind but without any suggestion that the computer itself actually thinks, believes, and so on.

and:

> weak AI *sim.-non-human* wherein the computer system is in some sense purported to be a simulation of the cognitive processes of a non-human mind (say the mind of a bat, frog, or armadillo) but without any suggestion that such processes are literally instantiated in the computer.

And weak AI should also subsume what prior to the early 1970s was often simply called Artificial Intelligence, or what I would now wish to call

> weak AI *task non-sim.* wherein the computer system is designed so that it can perform tasks that hitherto required intelligence, but *no* intelligence is required of the machine; nor is the machine system to be viewed as a simulation of cognitive processes (human or other). (In these projects the term ''artificial'' is emphasized.)

It would be inaccurate to suggest that researchers in machine intelligence woke up one morning in the 1970s like a collective Monsieur Jourdain to find they had all been cognitive psychologists for a decade. But it is true, I think, that a "psychologizing" of the field developed as ambitions shifted from the aims of what I've labeled weak AI *sim. human* and weak AI *task non-sim.* to the more "heady" goals of strong AI *human*.

Adding to that atmosphere were remarks such as those made by Zenon Pylyshyn in his highly provocative article "Complexity and the Study of Human Intelligence" (1979):

> In spite of the apparent preoccupation with engineering, I believe that the field of AI is co-extensive with that of cognitive psychology.[7]

And somewhat earlier Allen Newell and Herbert A. Simon in "Computer Science as Empirical Inquiry" (1976) had claimed:

> The study of logic and computers has revealed to us that intelligence resides in physical-symbol systems. This is computer science's most basic law of qualitative structure.[8]

And the close connection (overlap?) they perceived between strong AI *human* and CS (or "weak" AI) *sim.-human* was put as follows:

> We can gain some perspective on what is going on by turning, again, to the analogy of the germ theory. If the first burst of research stimulated by that theory consisted largely in finding the germ to go with each disease, subsequent effort turned to learning what a germ was—to building on the basic qualitative law a new level of structure. In artificial intelligence, an initial burst of activity aimed at building intelligent programs for a wide variety of almost randomly selected tasks is giving way to more sharply targeted research aimed at understanding the common mechanisms of such systems.[9]

[7] Reprinted in Haugeland, *op. cit.*, pp. 68–69.
[8] Reprinted in Haugeland, *op. cit.*, p. 64.

And under their next heading,"The Modeling of Human Symbolic Behavior," they write:

> The symbol system hypothesis implies that the symbolic behavior of man arises because he has the characteristics of a physical symbol system. Hence, the results of efforts to model human behavior with symbol systems become an important part of the evidence for the hypothesis, and research in artificial intelligence goes on in close collaboration with research in information processing psychology, as it is usually called.[10]

There remains, however, an ambiguity in their characterization of AI. For it is unclear whether "intelligent programs" are to be viewed as programs that would literally *impart* intelligence to a machine, or if instead they are to be interpreted as programs that provide a symbol system model that *depicts* human intelligence without actually partaking of it. Sometimes this more cautious rendering seems the most accurate (and sympathetic) way of paraphrasing their position; sometimes not.

It should be noted that what I have called weak AI *task non-sim.* may not be as uncontroversial a category as it first seems. In fact, weak AI *task non-sim.* may not even be possible if Zenon Pylyshyn is correct when he writes:

> any AI system is at some level a psychological theory, simply because the description of the intelligent task to which it is addressed already is essentially a description of some psychological process.[11]

Part of Pylyshyn's reason for claiming this derives from his belief that the distinction between *what* an organism does and *how* it does it is fuzzier than at first appears, and that any satisfactory description of what a person (or machine) does will go beyond "a summary of the regular-

[9] *Op. cit.*, pp. 48–49.
[10] *Loc. cit.*
[11] *Op. cit*, p. 85.

ities in some corpus of behavior'' and include theoretical commitments involving generalizations about the behavior.

His remarks on what might be dubbed the ''mix'' between the *how* and *what* of executing tasks strike me as interesting and important. But it is imperative to see that they cannot be converted into arguments against the possibility of what I have called weak AI *task non-sim*. Nor do I mean to suggest that Pylyshyn has said anything that would imply this. What I wish to stress and what I think has been increasingly overlooked is that any adequate taxonomy of the field of machine intelligence must leave room for a category that tolerates the possibility of machines performing tasks that had hitherto required intelligence (human or other) and *performing them in ways totally bereft of any psychological reality*. This is a more elaborate way of making the point illustrated at the end of Chapter Two with the analogy of a steam drill digging railway tunnels *without* using muscles (i.e., no muscular reality in that steam drill!).

It is all too easy to think that if a machine is capable of performing a type of task that previously required human intelligence, *then* the machine should be credited with at least a little bit of intelligence—say, just at the task level (whatever that might mean!). But if such an inference were warranted, it should also be true that if a human being were capable of performing a type of task using intelligence that had hitherto seemed performed by a machine without intelligence, then *either* (a) the machine should in retrospect be interpreted as also having been at least a little bit intelligent all along, *or* (b) the human being's performance of the task should be reinterpreted as not having really required intelligence. But consider some broad range of what might be called ''machine indigenous'' tasks—tasks that *only* came into being when certain machines were developed to perform them—such as gumballs displayed within a glass sphere being dispensed down a shoot whenever a nickle is deposited in the machine's slot.[12] Now we could imagine installing a human being in a glass sphere containing gumballs; through the exercise of intelligence (not much, just a little), that human being could dispense gumballs down a shoot upon the deposit of a nickle in a slot. Obviously, from the fact that we could "psychologize" such a "machine indigenous" task, it would not follow that ordinary gumball machines prior to that, *or* after that (*the chronological order is irrelevant*) should *thereby* be credited with intelligence in any degree. Nor, of course, would there be sufficient reason to subtract intelligence from what we should ascribe to our humanized gumball dispenser. Consequently, however "strong" AI is to be construed, its domain should not be so broad as to include as examples all machines that performed tasks that have been or could be performed by human beings through the exercise of their (human) intelligence. For if this were so, *any* machine executing "machine indigenous" tasks would also automatically qualify as an AI machine the minute someone conjured up a way for a human being to execute the type of task in question utilizing intelligence!

Part of the utility in having (at least) as many categories for both strong and weak AI as proposed above is that they can assist in pinpointing what the actual psychological (or philosophical) upshot of a given project is, regardless of how those projects may be described by the researchers who devise them. It can be tempting to dismiss a given project as without psychological (or philosophical) signifi-

[12] A task performed by a machine in *a radically different way* from the way it has typically been performed by human beings could also count as a "machine indigenous" task, as I am using the phrase. For the most part, though, I mean it as a label for types of tasks that had never been performed in any way at all prior to some machine's execution of them.

cance when it fails to "live" up to its perpetrator's claims; as, for example, a project successful at strong AI *human*. Suppose such claims are indefensable, as so far they have all seemed to be. It is still relevant to ask whether they might nevertheless be of interest psychologically (or philosophically) if categorized in some different way: as, for example, a project in strong AI *non-human* or weak AI *sim.-human*.

As instances, neither Winograd's much discussed program SHRDLU (1972)[13] for natural language comprehension—a block-manipulating, robot-arm simulation—nor Schank's well-publicized programs for story comprehension (1975)[14] seem to achieve the goals of strong AI *human*. While ignoring the details, let me simply record that Dreyfus's critique of both, in his "From Micro-Worlds to Knowledge Representation" (1979, 1981)[15] and Searle's attack directed at Schank but designed to be applicable to Winograd's work as well as others, in his "Minds, Brains, and Programs" (1980), impress me as quite convincing in many respects, and at least convincing enough to show that whatever the theoretical payoff from those projects may be, it is not that they have arrived at the El Dorado of psychological modeling, namely, strong AI *human*. Nor do I think that any of the problem-solving simulation programs of Simon, Shaw, and Newell can be viewed as successful at strong AI *human*. But it is still pertinent to ask how suggestive any of them are when viewed as projects in weak AI *sim.-human*. My own

[13]Terry Winograd, "Understanding Natural Language," *Cognitive Psychology*, 1 (1972), pp. 1-191; also published as a book (New York, 1972).

[14]Roger C. Schank, "The Primitive Acts of Conceptual Dependency" and "Using Knowledge to Understand" in *Theoretical Issues in Natural Language Processing* (Cambridge, Massachusetts, June 10-13, 1975).

[15]Reprinted in Haugeland, *op.cit.*, pp. 161-204.

answer to this in Chapter Four in connection with the earlier work of Simon, Shaw, and Newell was a somewhat mixed "not very." And a rather similar appraisal of Kenneth Colby's PARRY, developed in his "Modeling a Paranoid Mind" (1981),[16] was argued for in my comments on that project in "Paranoia Concerning Program-Resistant Aspects of the Mind, and Let's Drop Rocks on Turing's Toes Again" (1981).[17] And I think it is a fair summary of Dreyfus's position in "From Micro-Worlds to Knowledge Representation" to say that he believes that both Winograd's program and Schank's, as well as numerous others, fail at even the psychologically less robust goals of weak AI *sim.-human*. In this respect it might be noted that the intended scope of Dreyfus's criticisms of machine intelligence projects seems broader than Searle's, which is explicitly aimed at strong AI. Whether Searle's objections should be viewed as only applicable to the species of what I have called strong AI *human* or include as well strong AI *non-human* seems to be unclear and interestingly so. That would depend in part on whether any truly intelligent being would have to possess intentionality, or whether there could be some more "offbeat" (to us) type of intelligence devoid of it. The mind-boggling nature of minds and their possible varieties, and the radical differences of some compared with our own, are nicely imagined in Paul M. Churchland's chapter "Expanding our Perspective" in *Matter and Consciousness* (1984), where he writes:

> Strange sense organs aside, the particular cluster of cognitive talents found in us need not characterize an alien species. For example, it is possible to be highly intelligent and yet lack all capacity for manipulating numbers, even the ability to count past five. It is equally possible to be highly intelligent and yet

[16] *The Behavioral and Brain Sciences*, 4, No. 4 (1981), pp. 515–60.
[17] *The Behavioral and Brain Sciences*, 4, No. 4 (1981), pp. 537–39.

> lack any capacity for understanding or manipulating language. Such isolated deficits occasionally occur in humans of otherwise exemplary mental talents. The first is a rare but familiar syndrome called *acalculia*. The second, more common, affliction is called *global aphasia*. We must not expect, therefore, that a highly intelligent alien species must inevitably know the laws of arithmetic, or be able to learn a system like language, or have any inkling that these things even exist. These reflections suggest further that there may be fundamental cognitive abilities of whose existence *we* are totally unaware!
>
> Finally we must not expect that the goals or concerns of an alien intelligent species will resemble our own, or even be intelligible to us. The consuming aim of an entire species might be to finish composing the indefinitely long magnetic symphony begun by their prehistoric ancestors, a symphony where the young are socialized by learning to sing its earlier movements. A different species might have a singular devotion to the pursuit of higher mathematics, and their activities might make as much sense to us as the activities of a university mathematics department would make to a Neanderthal. Equally important, racial goals themselves undergo evolutionary change, either genetic or cultural. The dominant goals of our own species, 5,000 years hence, may bear no relation to our current concerns. All of which means that we cannot expect an intelligent alien species to share the enthusiasms and concerns that characterize our own fleeting culture.[18]

It could be argued that what some AI projects teach us is that some species of intelligence such as calculation or computation, which includes checker playing, theorem proving, and the like, are detachable from more high-powered psychological notions such as understanding and can exist without it or other ''Brententional'' accessories, even though in our own case they are constantly accompanied by them. (The issue of the interconnectedness or

[18](Cambridge, Massachusetts), p. 156.

detachability of mental notions with or from one another will be examined later.)

A somewhat precious but tantalizing note: if it makes sense—as it certainly seems to—to allow for the possibility of strong AI *non-human* (*machine*), where the intelligence ascribable to the machine is peculiar to it (or its type), it also makes sense to allow for the possibility of weak AI *sim.-non-human* (*machine*). In other words, an AI project devoted to the *simulation* of "machine indigenous" intelligence! The intriguing question arises as to whether such machine simulations of "machine indigenous" minds would be perforce without point, since we were the makers of the mindy (strong AI) artifacts to begin with.[19] (Why on earth bother with weak AI *sim.-non-human* [*machine*]!?) Or instead would we view those artifacts in the way we sometimes view other items (such as literature or painting) of which we are the demiurge creators, namely, with awe and incomprehension, sometimes to the point of wanting to devise other theories to aid in our understanding of them? (Consider: the need[?] for literary criticism.) Here we can imagine an eventual reflective robotology standing to a rapidly expanding, richly varied, and psychologically exciting yet in many ways puzzling historical world corpus of strong AI projects involving individual talents, traditions, genres, movements, and so on—much in the way that literary criticism and theory stands to a puzzling historical world corpus of poems, plays, novels, and so on. And within this context the otherwise redundant-seeming category "weak" AI *sim.-nonhuman* (*machine*) could take on a theoretical and critical utility. (See the remark on experimental aircraft with "bugs" in them, mentioned at the end of the Epilogue.)

[19] See my "Purposes and Poetry," in *Body, Mind, and Method*, eds. D. F. Gustafson and B. L. Tapscott (The Hague, 1979) pp. 203–24; especially p. 203.

The Question "Can Machines Think?" and the Conceptual Space between "Yes" and "No": Digressions on Locke and Animal Automatism, Thinking and "Thinking," Haugeland's "Derivative Intentionality," Dennett's "Intentional Stance," plus Other Excitements

A striking aspect of the philosophical debates generated by Descartes' doctrine of animal automatism was the calcification of perspectives on the matter: namely, that animals were viewed either as our full-fledged cognitive affective equals (or betters!) or as mental vacants. It was with Locke's *Essay Concerning Human Understanding* (1690) that a highly influential, more tempered, though dramatically less fur-raising alternative was outlined: namely, that animals shared with us the ability to receive, discern, compare, and compound various ideas (perceptions) but differed from us in lacking the capacity for forming general abstract ideas and annexing names to them. For present purposes, what seems instructive about Locke's position is not its categories or details but rather its orientation to the entire controversy. Locke's answer to the question "Can animals think?" was neither the resounding "Yes" of a Montaigne, Charron, or La Mettrie (as discussed in Chapter One), nor the species-chauvinistic "No" of a Descartes or Malebranche. It was rather the more boring, yet, I think, much more obviously plausible, "To some extent." In a parallel way contemporary skirmishes concerning AI, and philosophical debates about minds and machines in general, seem either to sidestep altogether or to downplay the significance of the range of possible "To some extent" (or "Sort of") answers to the question "Could (or do) machines think?"

The nomenclature of the previous section is designed in part to illustrate some of the variety of ways and degrees to which machines might be viewed as having psychological (and philosophical) significance. Flexibility even at a mere terminological level can be helpful in mitigating against the rigidity and polarization of positions that has frequently attended contemporary discussions of robot psychology from the appearance of Turing's "Computing Machinery and Intelligence" on into the present.

I think it unquestionable that there exists a proverbial gray yet interesting conceptual space between the "Yes" of the more evangelical proponents of "strong" AI and the intrepid "No" of their opponents. And furthermore I suspect that it is somewhere within that space that machine-intelligence projects and a developing robotology should receive their most reasonable characterizations for many years to come. Consider the following analogy: Joe falls madly in love with Martha and commences courting her. After a few weeks someone asks Joe how it is going with Martha, to which Joe replies that although he is madly in love with her, she at most tolerates him. Nevertheless, he persists. After a few more weeks the same interrogator repeats the question and Joe, smiling slightly, says he thinks she is beginning to like him. Still later Joe is queried a third time, and this time with beaming visage he says that although she may not love him, he thinks she "loves" him.

Now admittedly love is different from "love"; nevertheless, who is to say that the transformation from likes to "loves" in the affections of Martha for Joe does not represent *an important development* for Joe? So, too, why should there not be room for a type of transformation in technology from machines that exhibited *no* capacity for thinking to those that we felt forced to describe as "thinking," a transformation which heralded an important theoretical advance for those interested in a mechanistic modeling of the mind?

There is, of course, the crucial problem of how to sort out from the myriad of possible scare-quote cases of machines or robots "thinking," "intending," "feeling," etc., those that are actually psychologically illuminating or suggestive. And obviously there is no discovery procedure for doing that. But intuitively, at least, it seems that interesting cases of "thinking" might attend some of the machines *cum* programs that could be plugged into the categories weak AI *sim.-human* and weak AI *sim.-non-human*. It could even come about that what we might initially regard as a psychologically *uninteresting* case of a machine "thinking" would after further analysis be redescribed as "strong" AI *non-human* (for example, a case of "machine indigenous" thought). In other words, "thinking" or any other scare-quoted mental act could have a kind of quasi-theoretical interim use for tagging those innovative instances of machines or robots about which we are not at first sure what to say.

If we should ever be treated to instances of "machine indigenous" thinking, there is no reason to suppose that we would immediately be at ease with them conceptually or even recognize them for what they were. For even if they were explicable in the sense that their designers and programmers could reconstruct for us the nature of their creations, such artifacts might still puzzle us with respect to how they were similar to or different from cases of cognitive-affective systems more familiar to us such as ourselves and other animals. And if we remind ourselves that what is more familiar is not necessarily better understood—for example, our own minds!—the difficulties multiply. What we could be faced with is a case of having to construct as well as decipher comparisons between unfamiliar technological novelties and familiar but dimly understood psycho-biological antiquities (ourselves). That such comparisons would become comprehensible simultaneously with the advent of any technological breakthrough is, I think, something no one has any particular reason to expect. (Compare: For years, without much success, I operate a small firm that manufactures gizmos which I hope will eventually have a wide appeal and make me a lot of money. Then during a month when I am flirting with some new marketing techniques, my gizmos "take off," develop a wide appeal, and make me a lot of money. Although I may, in a sense, know exactly what I have done, there is also the possibility that in a deep sense I fail to understand what's happened.)

The possibility of ascribing scare-quotes thinking (rather "thinking") to machines in a manner that suggested that a psychology exists for them has a bearing on some issues raised in John Haugeland's interesting introduction to his anthology *Mind Design* (1981).[20]

Haugeland discusses reservations one might have to cybernetically oriented cognitive psychology when conceived in a certain way: namely, as a theory based on what, following Daniel Dennett, is called a "semantic engine" or an "automatic formal system with an interpretation such that the semantics will take care of itself." Haugeland writes:

> The discovery that semantic engines are possible—that with the right kind of formal system and interpretation, a machine can handle meanings—is the basic inspiration of cognitive science and artificial intelligence. Needless to say, however, mathematics and logic constitute a very narrow and specialized sample of general intelligence. People are both less and much more than automatic truth-preservers.[21]

He later claims:

> The basic idea of cognitive science is that *intelligent beings are semantic engines*—in other words, automatic formal systems with interpretations under which they consistently make sense. We can now see why this includes psychology and artificial intelligence on a more or less equal footing: people and

[20] Haugeland, *op. cit.*
[21] *Op. cit.*, p. 24.

> intelligent computers (if and when there are any) turn out to be merely different manifestations of the same underlying phenomenon. Moreover, with universal hardware, *any* semantic engine can in principle be formally imitated by a computer if only the right program can be found.[22]

And then anticipates objections to such an interpretation of psychological science:

> Of course, it is possible that this is all wrong. It might be that people just *aren't* semantic engines, or even that no semantic engine (in a robot, say) can be genuinely intelligent. There are two quite different strategies for arguing that cognitive science is basiclly misconceived. The first, or *hollow shell* strategy has the following form: no matter how well a (mere) semantic engine acts *as if* it understands, etc., it can't *really* understand anything, because it isn't (or hasn't got) "X" (for some "X"). In other words, a robot based on a semantic engine would still be a sham and a fake, no matter how "good" it got.[23]

One of the "X"s he considers as a candidate for what a semantic engine could never have is original intentionality. He writes:

> A different candidate for "X" is what we might call *original intentionality*. The idea is that a semantic engine's tokens only have meaning because we give it to them; their intentionality, like that of smoke signals and writing, is essentially borrowed, hence *derivative*. To put it bluntly: computers themselves don't mean anything by their tokens (any more than books do)—they only mean what we say they do. Genuine understanding, on the other hand, is intentional "in its own right" and not derivatively from something else. But this raises a question similar to the last one: What does it take to have original intentionality (and how do we know computers can't have it)? If we set aside divine inspiration (and other

[22] *Op. cit.*, p. 31.
[23] *Op. cit.*, pp. 31–32.

> magical answers), it seems that original intentionality must depend on whether the object has a suitable structure and/or dispositions, relative to the environment. But it is hard to see how these could fail to be suitable (whatever exactly that is) if the object (semantic engine *cum* robot) always *acts* intelligent, self-motivated, responsive to questions and challenges, and so on. A book, for instance, pays no attention to what "it" says—and that's (one reason) why we really do not think it is the *book* which is saying anything (but rather the author). A perfect robot, however, would seem to act on its opinions, defend them from attack, and modify them when confronted with counter-evidence—all of which would suggest that they really are the *robot's own* opinions.[24]

Haugeland's reservations about the argument he has aptly reconstructed seem to me well taken, and complement the type of reasons given in Chapter Three for being suspicious of the defenders of La Mettrie's watch. Moreover, I suspect there are indefinitely many possible compromise candidates conceptually situated between "derivative intentionality" and Brentano's non-derivative kind. Owing to their nature as at least quasi-self-adaptive systems that perform tasks and utilize relatively rich, changeable, informational processing procedures, various robots and machines including "semantic engines," even though they fall short of mimicking us in terms of "original intentionality," could still be in interesting systematic respects psychologically "deeper" than books and smoke signals, which are in no sense active *manipulators* of their meanings.

Something not unlike the just expressed attitudes toward the possibility of ascribing "original intentionality" to a cybernetic system seems to underlie Daniel Dennett's intentionalist approach to the philosophy of mind, which was first propounded in his *Content and Consciousness* (1969) and later elaborated in *Brainstorms* (1978).

[24] *Op. cit.*, pp. 32–33.

In the former work he was concerned with the well-known problem of being able to show (a) how mental phenomena that *seem different* from physical phenomena by virtue of their having a content or being about something (which may or may not exist) really are reducible to physical phenomena; *or* (b) why, even if they are not so reducible, this is of minor metaphysical moment; *or* (c) how, given the failure of showing (a) or (b), one can nevertheless admit the subject matter of intentionality into the framework of science in general. Dennett's own way of stating the problem was this:

> The first challenge is the irreducibility hypothesis, that the Intentional cannot be reduced to the non-Intentional, or, as we have seen, the extensional. Then the evidence comes in that we can neither do without the Intentional, or cleave to it alone, for there are signs that the possibility is remote of a successful non-Intentional behaviourist psychology; and the alternative of an entirely Intentional psychology would entail a catastrophic rearrangement of science in general.[25]

And his own strategy for coping with the problem is proposed in the following remarks:

> Fortunately, however, once the problem of Intentionality is clearly expressed, it points to its own solution. There is a loophole. The weak place in the argument is the open-endedness of the arguments that no extensional reduction of Intentional sentences is possible. The arguments all hinged on the lack of theoretically reliable *overt* behavioural clues for the ascription of Intentional expressions, but this leaves room for *covert*, internal events serving as the conditions of ascription.[26]

He then writes:

> Could there be a system of internal states or events, the exten-

[25] Dennett, *op. cit.*, p. 39.
[26] *Loc. cit.*

> sional description of which could be upgraded into an Intentional description? The answer to this question is not at all obvious, but there are some promising hints that the answer is Yes.[27]

Later, in *Brainstorms*, in his chapter "Intentional Systems" I take Dennett to be developing in more detail what such "upgrading" might involve. For Dennett, a system is intentional "only in relation to the strategies of someone who is trying to explain and predict its behavior." We are said to adopt an "intentional stance" toward a system when we make assumptions of rationality about it and attempt to predict its behavior by "ascribing to the system *the possession of certain information* and supposing it to be *directed by certain goals*" and then "by working out the most reasonable or appropriate action on the basis of these ascriptions and suppositions." He considers our adopting just such an intentional stance vis-à-vis a chess-playing computer and claims:

> It is a small step to calling the information possessed the computer's *beliefs*, its goals and subgoals its *desires*. What I mean by saying that this is a small step, is that the notion of possession of information or misinformation is just as intentional a notion as that of belief. The "possession" at issue is hardly the bland and innocent notion of storage one might suppose; it is, and must be, "epistemic possession"—an analogue of belief. Consider: the Frenchman who possesses the *Encyclopedia Britannica* but knows no English might be said to "possess" the information in it, but if there is such a sense of possession, it is not strong enough to serve as the sort of possession the computer must be supposed to enjoy, relative to the information it *uses* in "choosing" a chess move. In a similar way, the goals of a goal-directed computer must be specified intentionally, just like desires.
>
> Lingering doubts about whether the chess-playing computer *really* has beliefs and desires are misplaced; for the definition

[27] *Op. cit.*, p. 40

> of intentional systems I have given does not say that intentional systems *really* have beliefs and desires, but that one can explain and predict their behavior by *ascribing* beliefs and desires to them, and whether one calls what one ascribes to the computer beliefs or belief-analogues or information complexes or intentional whatnots makes no difference to the nature of the calculation one makes on the basis of the ascriptions.[28]

And

> We do quite successfully treat these computers as intentional systems, and we do this independently of any considerations about what substance they are composed of, their origin, their position or lack of position in the community of moral agents, their consciousness or self-consciousness, or the determinacy or indeterminacy of their operations.[29]

I quote Dennett at length because of the dexterous conceptual juggling act the foregoing passages display: he is not willing to commit himself to saying "intentional systems *really* have beliefs and desires to them." Then there is his seeming indifference as to whether what one ascribes to them is treated as beliefs or belief-analogues, and his claim that he does not mean to imply that the chess-playing computer he imagines would be "a *completely* [my italics] adequate model or simulation of Mind." Added to this is his view that ascribing intentionality to such computers would be pragmatic and not "intrinsically right or wrong."

This up-in-the-air indefiniteness about how much psychology Dennett's imagined chess-playing computer is supposed to partake of is reflective of an interesting recurrent ambiguous aspect of research in machine intelligence and the philosophizing about it. Something like this: (1) machines can handle information or meanings in ways sig-

[28] Reprinted in Haugeland, *op. cit.*, pp. 224–25.
[29] *Op cit*, p. 225.

nificantly different from the ways in which smoke signals or books can be viewed as bearers of meaning, yet (2) there is also a difference between any such cybernetic systems and human beings which no doubt has something to do with consciousness in the latter and the lack of it in the former. So in the case of intentionality, for example, there is a sense in which machines seem to have it and a sense in which they do not. This could be called the "To have and have not" (or Bogart/Bacall) aspect of AI. If one is more enamored of the "have" than the "have not" half of the situation, one is likely to be a proponent of strong AI in one form or another. If, however, one is instead smitten by the "have not" feature, one is likely to believe that projects in machine intelligence are at best projects in some form of weak AI. If, however, one is impressed with both aspects, a more ambivalent stance—such as that adopted by Dennett—is likely to result.

A good example of a philosophical stance toward machine intelligence that is sensitive to degrees of cognitive difference across various projects and that avoids simple "Yes" or "No" answers to the question "Can machines think?" is provided by Churchland in his comparisons between Joseph Weizenbaum's program ELIZA, which simulates a psychotherapist in conversation carried out by typewriter with a "patient," and Winograd's block world item organizer SHRDLU. Churchland quotes the following exchange between "patient" and computer-psychotherapist contained in Weizenbaum's *Computer Power and Human Reason*:[30]

> Men are all alike.
> IN WHAT WAY
> They're always bugging us about something or other.
> CAN YOU THINK OF A SPECIFIC EXAMPLE
> Well, my boyfriend made me come here.

[30] (San Francisco, 1976). And in Churchland, *op.cit.*, pp. 117–18.

YOUR BOYFRIEND MADE YOU COME HERE
He says I'm depressed much of the time.
I AM SORRY TO HEAR YOU ARE DEPRESSED
It's true. I am unhappy.
DO YOU THINK COMING HERE WILL HELP YOU NOT TO BE UNHAPPY
I need some help, that much seems certain.
WHAT WOULD IT MEAN TO YOU IF YOU GOT SOME HELP
Perhaps I could learn to get along with my mother.
TELL ME MORE ABOUT YOUR FAMILY
My mother takes care of me.
WHO ELSE IN YOUR FAMILY TAKES CARE OF YOU
My father.
YOUR FATHER
You are like my father in some ways.
WHAT RESEMBLANCE DO YOU SEE
You are not very aggressive, but I think you don't want me to notice that.
WHAT MAKES YOU THINK I AM NOT VERY AGGRESSIVE
You don't argue with me.
WHY DO YOU THINK I DONT ARGUE WITH YOU
You are afraid of me.
DOES IT PLEASE YOU TO BELIEVE I AM AFRAID OF YOU
My father is afraid of everybody.
WHAT ELSE COMES TO MIND WHEN YOU THINK OF YOUR FATHER
Bullies.
DOES THAT HAVE ANYTHING TO DO WITH THE FACT THAT YOUR BOYFRIEND MADE YOU COME HERE

Churchland's assessment of ELIZA is as follows:

> In this conversation, the computer, running ELIZA, gives the eerie impression of real understanding. But in fact it has none. The responses are largely constructed from the patient's own

> sentences, by simple transformation, and from a stock of standard question forms tailored to key words from the patient (''depressed,'' ''like,'' and so on). ELIZA has no conception of what a father, a brother, or unhappiness is. It has no concept of these things, no understanding of what these words mean. Which just goes to show how surprisingly little understanding is required to engage successfully in many standard forms of conversation.[31]

But of Winograd's SHRDLU, which he regards as a ''much more impressive program,'' he writes:

> Its syntax is very sophisticated, and the program embodies some systematic information about the properties of the bodies that inhabit its world. Crudely, it knows, a little, what it is talking about. As a result, it can draw useful inferences and divine real relations, a talent that is reflected in the much more complex and keenly focused conversations one can have with it. Conversations must be restricted to the block world, however, and to those narrow aspects it encompasses. SHRDLU does not have an empty knowledge base, but its base is still less than microscopic compared with our own.[32]

Even though, as Churchland contends, the knowledge base of SHRDLU is ''microscopic compared with our own,'' one can argue, as he does, that *it* is impressive as compared to ELIZA. Even knowing ''a little,'' ''drawing useful inferences,'' and being able to ''divine real relations'' marks, I think, a philosophically and psychologically significant contrast between the two programs, a contrast which can be easily neglected if one is prone to the more Cartesian-type ''Yes'' or ''No'' posture toward cognition in machines. (Note: both ELIZA and SHRDLU could pass Turing-type tests, which, I think, only shows how ineffectual such tests are for showing the comparative psycholog-

[31] *Op. cit.*, p. 118.
[32] *Loc. cit.*

ical significance of different AI programs. More of this later.)

What I have called the conceptual space between "Yes and "No" answers to the question "Can machines think?" seems to me populated by an indefinitely rich heterogeneous set of "Sort of" answers. But I would like to make special mention of two different broad kinds of differences that can exist between human cognitive-affective capacities and performances and those attributable to other organisms and artifacts. (Some of these would be examples of a difference between thinks and "thinks," others not.) If, for example, we were to claim that a machine "thinks" in playing chess instead of thinks, what the scare-quotes *could* call attention to is that the machine's chess-playing performance failed to attain a certain *degree of alacrity* when compared with human levels of performance. Perhaps the machine's program included no effective end-game strategies, lacked other basic heuristics, was incapable of revising strategies that systematically led to failure, and so on. Let me label such a difference a *degree of alacrity difference*. A *degree of alacrity difference* is a difference that is, sometimes, *in fact*, erasable, and it is a difference that is almost always imaginable as erasable. The alacrity is considered as, in principle, improvable—e.g., refined end-game strategies in a chess-playing program—so that "thinks" sheds its scare-quotes and metamorphizes into thinks. The difference between "thinks" and thinks where that difference is a *degree of alacrity difference* I will call a *difference along a continuum*. It is not that the machine is playing chess in some qualitatively radically different way from the way thinking humans play chess. Rather, it is that given the deficiencies of their programs, they can only bungle along at the game at a lower level of performance, but performance of recognizably the same type. It is like the case where someone is learnig to be a cook for a restaurant but is not yet trusted to execute any main-dish recipes from the menu. Instead the person is limited to turning the bratwurst, tossing salads, folding strudel, and checking the consistency of the broth for the lentil soup. All these rather *basic* activities will be relevant to the apprentice chef's next stage of instruction and will be part of what is built on in the development of any further culinary art. The person is at most a ''cook,'' though being a ''cook'' may be continuous with becoming a cook.

Now a *degree of alacrity difference* can be contrasted with another kind of difference that is not a *difference along a continuum*, but one that might be called a *modus vivendi difference*.

By a *modus vivendi difference* I mean a difference rather like the differences we can detect among the following sorts of cases. Consider swimming. Many different kinds of creatures can swim: fish, turtles, shrimp, porpoises, and Mark Spitz. But none of them swim in quite the same way. How and why they swim is unique to how they live, move, and have their being, or what I will call, stretching the phrase a bit, their *modus vivendi*. It is not that their different capacities for swimming lie on a continuum in the way, say, the swimming of Mark Spitz and a little boy learning to dogpaddle lie on the same continuum, or in the way an apprentice chef's ''cooking'' lies on a continuum with a chef's cooking. In the case of the little boy learning to dogpaddle (two or three yards at a time), we might even say that the child at best ''swims,'' but that as he develops he *could*, perhaps, even come to attain the sorts of aquatic skills that brought Mark Spitz seven gold medals in the 1972 Olympics. In other words, the difference between the little boy's ''swimming'' and Mark Spitz's swimming is a *degree of alacrity difference*. But neither the boy (nor Mark Spitz) could develop further skills such that either would swim literally *like* a fish, turtle, shrimp, or porpoise. The difference in swimming between the two and any of the other marine life mentioned—or what involves a *difference across a spectrum*—I am calling a *modus vivendi difference*. Note that typically a *modus vivendi difference* between say, a human X-ing and some other sort of subject X-ing would not be indicated by scare-quotes. Fish, turtles, shrimp, porpoises, and Mark Spitz all swim, but in different ways. None are said only to "swim."

There is, of course, the ineluctable problem of borderline cases, and for some examples it will be difficult to decide exactly how the difference between a human X-ing and a machine X-ing should be expressed. Consider the case of a computer that plays checkers using only algorithms. Humans do not play checkers that way but utilize a variety of non-"fail-safe" heuristics. Should we describe the difference between the checker playing of the computer and that of the humans as a *modus vivendi* difference, or as a *degree of alacrity* difference? Is a basically pan-algorithmic checker-playing machine different in *kind* from an essentially heuristic human checker player? Or given the rather notorious blurry line between algorithms and heuristics, is the difference to be viewed primarily as a *degree of alacrity* difference?

In spite of borderline cases, however, I want to suggest that with respect to how a machine or robot does anything—plays checkers, composes poetry, orders hamburgers, etc.—and the difference between how that is done and how a human being might perform the same sorts of tasks constitutes either a *degree of alacrity difference along a continuum* or a *modus vivendi difference across a spectrum*, and that being able to say which sort of difference it is might be theoretically useful. It could, for example, enable one to size up a given machine-intelligence project as one in strong AI *non-human* as opposed to one aimed at strong AI *human* or weak AI *sim-human*, and so on.

One way to characterize some of the disputes between those who believe in the possibility of AI as a model of human mentality and those who would deny it is to see the former as believing that the difference between various machines that now "think" and ourselves who think is only a *degree of alacrity difference along a continuum*, erasable in principle degree by degree. The latter, on the other hand, would contend that although machines can compute, recognize patterns, prove theorems, play chess, etc., the difference between how they do any of these things and how we do them constitutes a *modus vivendi difference across a spectrum*, a kind of basic categoreal difference and certainly not a difference susceptible to being chipped away at by degrees. Here, note, the dispute is not just a dispute about whether we could build machines that think, but whether, even if we could, they would serve as models of our own mentality. I see no reason to suppose that there could not be machine indigenous mentalities that shed virtually no light on our own except for suggesting how different our minds are from "theirs." (Here I am reminded of a remark by Hilary Putnam (in conversation long ago) that although computers are not as "deep" as we are, the frightening thing is that they are so much faster!)

Many of the disputes in the seventeenth century between Descartes and Henry More et al. concerning animal intelligence might be redescribed by saying that Descartes denied that the difference between our own minds and the causes of animal behavior was anything like a *degree of alacrity difference*, and furthermore that the only *modus vivendi differences* between minds that existed were the differences between our minds and the minds of God and his angels. La Mettrie, on the other hand, clearly viewed the diferences between ourselves and apes as a *degree of alacrity difference along a continuum*—a difference conceivably erasable through changing the environment and tampering with the physiology (in particular, the eustachian canal) of the ape. And, in a very general way, David Premack's investigation of the chimpanzee mind (which he called "a furry little computer")[33] can, I think, be seen as in the tradition and spirit of La Mettrie: an investigation predicated on the belief that many of the interesting basic differences between our own minds and the minds of chimpanzees are *differences along a continuum* (though not thereby obviously erasable!). For example, see his *Intelligence in Ape and Man* (1976)[34] and "Does the Chimpanzee Have a Theory of Mind?" (with Guy Woodruff) (1978).[35]

Although the different differences described above were mentioned in order to help "taxonomize" some of the positions occupying the conceptual space between "Yes" and "No" answers to the question "Can machines think?" those differences by no means apply simply to comparisons between "thinking" and thinking, or different ways of thinking, but apply as well to mentalistic notions *simpliciter* (with or without scare-quotes) insofar as they are variously instantiated across both natural and artifactual species.

The Toe-Stepping Game and Searle's Chinese Room

John Searle's aforementioned "Minds, Brains, and

[33] At a conference on perception and cognition held at the University of Minnesota, David Premack prefaced his talk with something like the following remark: "You may wonder what I'm doing here at a conference with all these experts on cybernetics; well just think of the chimpanzee as a furry little computer." (Proceedings of this conference were published as *Perception and Cognition: Issues in the Foundations of Psychology*, Vol. IX of Minnesota Studies in the Philosophy of Science, ed. C. Wade Savage, Minneapolis, 1978.)

[34] Hillsdale, 1976.

[35] *The Behavioral and Brain Sciences*, 1, No. 4 (1978), pp. 515–26.

Programs'' seems to me one of the most important and highly controversial articles on minds and machines to be published during the last decade and a half. It, together with responses to it and Searle's responses to the responses, comprises a treasure trove of converging and diverging attitudes toward the possibility of machine intelligence within the framework of innovative computer programming. He considers the work of Roger Schank et al. in developing programs that enable machines to engage in question-and-answer behaviors concerning stories. In most respects, Schank's programs impart to machines capacities for performing Imitation Game-type tasks. Schank's machines are given stories and then asked questions about them, and the answers they provide to those questions seem to be pretty much the sort of answers we might imagine being made by human beings. Searle claims that the partisans of strong AI not only argue that such machines simulate human abilities, but also contend that

> (a) the machine can literally be said to *understand* the story and provide answers to questions; and (b) what the machine and its program do *explains* the human ability to understand the story and answer questions about it.[36]

Searle believes that neither (a) nor (b) are justified by Schank's work, and his basic argument runs as follows: Searle, who is totally ignorant of Chinese but fluent in English, is incarcerated in a room where he is given two batches of Chinese writing plus some rules in English for correlating them solely on the (formal) basis of their shapes. He is also given a third batch of Chinese inscriptions with further instructions in English for hooking up the third batch with the first two. The latter instructions also tell him how he is ''to give back certain Chinese symbols with certain sorts of shapes in response to certain sorts of shapes'' given to him in the third batch. The

[36] Searle, in Haugeland, *op. cit.*, p. 284.

people providing Searle with all the Chinese writing and the correlation rules in English refer to the first batch as "a script," the second batch as "a story," the third batch as "answers to the question," and the rules in English as "the program." Searle then imagines himself getting so good at following the instructions for manipulating the Chinese inscriptions and the programmers getting so good at concocting sets of instructions that Searle's "answers" to questions cannot be distinguished from native Chinese speakers. Furthermore, Searle is also provided stories written in English, about which he is asked questions, which he answers with an alacrity comparable to other native speakers of English. The upshot of it all is that although an external observer might view Searle's answers to the Chinese questions about the Chinese stories and his answers to the English questions about the English stories as equally adept, only in the latter case was there any real understanding involved. In the former case Searle claims he simply behaved like a computer. He was, in effect, "simply an instantiation of the computer program." He goes on to say:

> Now the claims made by strong AI are that the programmed computer understands the stories and that the program in some sense explains human understanding. But we are now in a position to examine these claims in light of our thought experiment.
>
> (a) As regards the first claim it seems to me obvious in the example that I do not understand a word of the Chinese stories. I have inputs and outputs that are indistinguishable from those of the native Chinese speaker, and I can have any formal program you like, but I still understand nothing. Schank's computer for the same reason understands nothing of any stories whether in Chinese, English, or whatever, since in the Chinese case the computer is me; and in cases where the computer is not me, the computer has nothing more than I have in the case where I understand nothing.

> (b) As regards the second claim—that the program explains human understanding—we can see that the computer and its program do not provide sufficient conditions of understanding, since the computer and the program are functioning and there is no understanding. But does it even provide a necessary condition or a significant contribution to understanding? One of the claims made by the supporters of strong AI is this: when I understand a story in English, what I am doing is exactly the same—or perhaps more of the same—as what I was doing in the case of manipulating the Chinese symbols. It is simply more formal symbol manipulation which distinguishes the case in English, where I do understand from the case in Chinese, where I don't. I have not demonstrated that this claim is false, but it would certainly appear an incredible claim in the example. Such plausibility as the claim has derives from the supposition that we can construct a program that will have the same inputs and outputs as native speakers, and in addition we assume that speakers have some level of description where they are also instantiations of a program.[37]

Given my objections to Turing based on the rock box in the toe-stepping game, it might seem obvious that I would feel considerable empathy with Searle's overall strategy. Recall his claim that

> I have inputs and outputs that are indistinguishable from those of the native Chinese speaker, and I can have any formal program you like, but I still understand nothing.[38]

So, too, with respect to the toe-stepping game, one could have infoots and outfoots indistinguishable from those in a pan-human toe-stepping game and any sophisticated rock box you like, and still not have genuine imitative behavior. The question arises—i.e., I have been asked this question by a number of philosophers, and also asked it of myself—as to how Searle's Chinese room argument against

[37] *Op. cit.*, pp. 285–86.
[38] *Op. cit.*, p. 285.

strong AI is like or unlike the rock box, toe-stepping game parody of Turing. The philosophical interest the answer holds seems to me to be this. The form of the argument contained in the toe-stepping game parody is incredibly elementary—almost crude—and transparently so, whereas Searle's argument at least initially sounds more sophisticated than that parody. Indeed, I think it *is* more sophisticated, at least didactically, because the major components of current AI projects are conspicuously mimicked within his example. But it also appears to contain more conceptual "folds," and it differs from the toe-stepping game in that part of its persuasiveness derives from focusing on the perspective of the person (Searle himself!) producing *both* the imagined outputs and his (Searle's) awareness that in the case of some of them (those in Chinese) no understanding of the language was involved, whereas in the other cases (involving English) it was. So one might think of the rock box example as a kind of "primitive" didactic ancestor to the Chinese room. However, an updated version of the toe-stepping game parody—a kind of "Chinese" toe-stepping game, if you like—is easily imagined. All one needs is to provide a human toe-stepper (e.g., Searle) with a rock box which he can manipulate so that sometimes the toe-stepping outputs he is responsible for are determined by his "blind obedience" to an instruction such as PUSH BUTTON ON ROCK BOX, whereas others involve him in genuine "imitative behavior" using his own foot. Then the arguments turn out to be exactly parallel, though it should be obvious that nothing of substantive significance was contained in the "Chinese" toe-stepping game situation which was not already present in the original parody. So now I tend to believe that in *most* respects, the toe-stepping game parody and the Chinese room argument stir-fry out into roughly the same course. Yet there is an important and interesting ingredient—more than a touch of five spice, dash of sesame oil—not detectable in the toe-stepping game serving, with which I think Searle flavors his polemic. For he claims that the force of his argument is not "simply that different machines can have the same input and output while operating on different formal principles." He writes:

> That is not the point at all—but rather that whatever purely formal principles you put into the computer will not be sufficient for understanding, since a human will be able to follow the formal principles without understanding anything, and no reason has been offered to suppose they are necessary or even contributory, since no reason has been given to suppose that when I understand English, I am operating with any formal program at all.[39]

But surely at least part of the point of Searle's Chinese room argument—and a crucial part—is that from similarities in inputs and outputs one cannot deduce similarity of intervening processes, states, or events. For a few paragraphs later, in his anticipation of what he calls "The Systems Reply (Berkeley)" to his argument, he says:

> Let us ask ourselves what is supposed to motivate the systems reply in the first place—that is, what *independent* grounds are there supposed to be for saying that the agent must have a subsystem within him which literally understands stories in Chinese: As far as I can tell the only grounds are that in the example I have the same input and output as native Chinese speakers and a program that goes from one to the other. But the point of the example has been to show that that couldn't be sufficient for understanding, in the sense in which I understand stories in English, because a person, hence the set of systems that go to make up a person, could have the right combination of input, output, and program and still not understand anything in the relevant literal sense in which I under-

[39] *Op. cit.*, pp. 286–87.

> stand English. The only motivation for saying there *must* be a subsystem in me which understands Chinese is that I have a program and I can pass the Turing test.[40]

So it at least appears that like the toe-stepping game parody, the Chinese room argument is designed to exhibit the inadequacy of Turing's Imitation Game test. Nevertheless it is also clear from the foregoing that there is an additional feature in Searle's polemic which he sees as central, and sometimes seems to emphasize to the point of overshadowing his explicit objections to the adequacy of the Turing Test. And this is his argument that purely formal programs are devoid of meaning or intentionality.

The rock box, toe-stepping game parody was aimed at illustrating in a cutting way how similar inputs and outputs would not guarantee similar intervening processes. But Searle has, at least in his gloss on his own argument, added to that sort of objection the further contention that the native English speaker who speaks no Chinese could even have the same intervening *program* between inputs and outputs as a native Chinese speaker and still understand nothing so far as Chinese was concerned. So the full-blown version of his primary objection to strong AI is not just that same inputs plus same outputs will not guarantee same intervening process (in this case one involving linguistic understanding), but that same inputs plus same outputs plus same intervening *program* will not guarantee same intervening process.[41] In other words, there is more to the intervening process leading to the production of

[40] *Op. cit.*, p. 290.

[41] Strictly speaking, this latter point seems to me independent of the Chinese room argument, although Searle treats it as part of that argument. In the end I do not think it matters much since it certainly complements the Chinese room strategy. Overall, Searle's article should be viewed as a rich interrelated cluster of forceful objections to strong AI. My emphasis on comparisons between the toe-stepping game and the Chinese room example is not meant to detract from that.

speaker-understood utterance or inscription-tokens than can be accounted for by a program—and, furthermore, there may not be any program involved at all! Meanings—intentionality, understanding—are necessary ingredients of *bona fide* intelligence, and no formal system is sufficient to guarantee the presence of them. Or as he has (in conversation) expressed it, "you can't get semantics from syntax." Obviously the point that a purely formal (syntactic) system does not in and of itself bestow meanings (much less some pragmatic demension) upon a machine that instantiated it is statable independently of a Chinese room (or rock box) type of example. On the other hand, who would ever have thought in the first place that formal systems such as computer programs might provide the Pygmalion one needed for modeling a mind unless there had been some sort of input-output Turing-type-test evidence to begin with!? That is, the charisma of the claims of strong AI depend initially on the dramatics provided by similarities in outputs and inputs between machines operating according to certain formal programs and human beings operating according to who-knows-what. And, the counter-charisma of Searle's Chinese room argument (or the toe-stepping game parody) depends on showing how inadequate Turing-type tests really are. Certainly Searle could have contented himself with arguments attempting to show that formal programs considered in and of themselves are mindless notations. But I think it is no exaggeration to say that the reverberations set off by his article depend on its first having taken issue with that amazingly persistent primary and perennial evaluation procedure used by researchers in machine intelligence over the last two decades, namely, Turing's Test.

In some respects Searle's strategy is reminiscent of Descartes' attack on Montaigne- and Charron-type arguments on behalf of animal intelligence. Recall (in Chapter One) that Descartes first aims to show that similar inputs and outputs are not sufficient to establish the existence of thought or understanding in an organism, and then proceeds to unmask what he thinks is really going on in animals which makes it possible for them to *appear* intelligent:

> They have no reason at all, and . . . it is nature which acts in them according to the disposition of their organs, just as a clock, which is only composed of wheels and weights is able to tell the hours and measure the time more correctly than we can do with all our wisdom.[42]

Wheels and weights in a clock, rocks rigged up in a rock box, and various formal programs in AI computers are just a few of the indefinitely many sorts of causal links between inputs and outputs insufficient for bestowing intelligence upon an artifact. The point Searle is nailing down, of course, *is that the entire research program of strong AI is predicated on the explicit but mistaken view that formal programs could be (or are) sufficient for the instantiation of intentionality or understanding in machines.*

My sympathy with Descartes' objections to arguments on behalf of animal intelligence based on input-output similarities and my commitment to the mindlessness of rock boxes predispose me to side with Searle in a general way against his critics. Nevertheless, I think the notion of a purely formal program—the distinction between syntax and semantics—or some other level of notational description involving intentionality *in some degree*—is not as clear-cut as he treats it. And for reasons that the previous section should have made obvious, his treatment of so-called derivative (or what he calls "metaphorical") intentionality seems to me problematic.

Searle writes:

> Our tools are extensions of our purposes, and so we find it natural to make metaphorical attributions of intentionality to

[42] Descartes, *Discourse*, *op. cit.*, p. 117.

> them; but I take it no philosophical ice is cut by such examples. The sense in which an automatic door "understands instructions" from its photoelectric cell is not at all the sense in which I understand English. If the sense in which Schank's programmed computers understand stories is supposed to be the metaphorical sense in which the door understands, and not the sense in which I understand English, the issue would not be worth discussing.[43]

I agree with him about the door—cf. its failure to be a candidate for polite or impolite behaviors in Chapter Two—but now have less settled views than I used to about whether all so-called purely formal programs would be as purely unmental as that. Again, a machine that "thinks" is not the same as a machine that thinks, but it also is not the same as a machine that neither thinks nor "thinks." Searle claims that "the computer understanding is not just (like my understanding of German) partial or incomplete; it is zero." On a scale of one to ten I would reserve zero for smoke signals or doors with electric eyes, and at least entertain the view that computers with programs such as Winograd's or Schank's might deserve a two or three. (Recall Churchland's remarks in the previous section on SHRDLU.)

With respect to Turing's Test, it should also be emphasized that not only is it inadequate as a test for there being literally a mind in a machine, it is for reasons given in Chapter Four inadequate as a test for a *simulation* in a machine of a mind. In this sense the chapter might be seen as trying to delineate how very weak Turing's Test actually is.

Three rather different but complementary objections to that test can be distinguished. The first is simply that input-output similarities between machines and cognitive-affective beings are not sufficient *or even necessary* for establishing interesting internal process, state, or event

[43] Searle, *op. cit.*, p. 288.

correlations. And it is these internal goings-on, I still believe (contra Ryle, various neo-Wittgensteinians, early Dennett (1971), et al.), to which at least a significant subportion of our mentalistic vocabulary somehow mysteriously refers and applies. To be content with Turing-type test criteria for assessing modeling success in strong AI human (or even weak AI sim.-human) is rather like thinking that slapping two halves of some analogical bun together makes a good model of a hamburger. ("Hey, where's the meat? the onion? Also, I wanted a pickle!") Conversely, to be discontent with a simulation model of cognitive processes because it failed to pass Turing's Test would be a mistake. For there *could* be an illuminating correspondence between internal processes, states, and events of some computer model and human processes, states, and events, yet deficient mapping at input and output levels. A cooked hamburger patty with onion and pickle upon it might give us a pretty good grasp of a hamburger even if an analogical bun were lacking and the whole repast were lodged within pita bread. ("Say, this tastes pretty much like a Big Mac!")

Second, as I argued in Chapter Four, a major drawback to using Turing-type tests for evaluating CS projects is that they fail to anticipate the problems posed by the possibility of many obviously psychologically non-equivalent programs being equally effective at enabling a machine to pass the test. This point is, I believe, similar in substance to Pylyshyn's claim in "Complexity and the Study of Artificial and Human Intelligence," where he writes:

> If we apply only the minimal constraint of "computing the same input-output function," then indefinitely many Turing machines would be equally viable "models."[44]

Furthermore, it is conceivable that one could have two psychologically non-equivalent machines one of which

[44] Pylyshyn, in Haugeland, *op. cit.*, p. 91.

passed Turing's Test and the other which did not, and where the latter comprised a more interesting model of human mentality than the former. (All these points could, of course, be reiterated in connection with projects in strong AI.)

A third objection to Turing's Test, which may be viewed as a special case of the first objection, is this. If there are program-resistant aspects of the mind, the criteria to be met in modeling these cannot be found along behavioral problem-solving lines. Turing's Test and all the modifications of it that I am aware of have with a vengeance been along just such lines. To model the symptoms or effects of our consciousness or "qualia" is not to model consciousness or "qualia." To model a pain in one's toe, it is not sufficient to model the tokening of some sentence such as "Ouch! my toe hurts." What would be sufficient for the modeling of program-resistant aspects of the mind, however, is itself a conceptually tangled topic which will be treated more fully in the final section. It is one thing to devise objections to Turing-type tests, and quite another to conceptualize interesting alternatives to them!

A Vacillation: The Spider Web Metaphor of the Interconnectedness of Mental Concepts vs. Condillac's Statue, or How Could Computation be as Singular as Rose Smell if the Mind Is Always Sort of All Over the Place!

In "The Imitation Game" objections to Turing's arguments on behalf of intelligence in computing machinery, I suggested that thinking was not a "monorail" type of activity that one or two examples could adequately characterize. It was rather, I claimed, a more all-purpose type of phenomenon. Compare: the Swish 600 and Descartes' view of reason as a "universal instrument" (Chapter

One). Underlying this interpretation of thinking was the intuitive feeling that terms such as 'thinks', 'reasons', 'believes', 'considers', 'intends' might be viewed as closely related; sort of like members of a linguistic clan that romp around together in the same connotative-denotative heather of semantic space. Where one shows up, the others do too. (Compare: "Uncle McDougall never visits, but he brings half the relatives with him.")

As is probably obvious, both Wittgenstein's *Philosophical Investigations* and Ryle's *Concept of Mind* for all their dissimilarities had an influence here. Each work in its own way was suggestive of a kind of "spread out" account of mentality, with psychological terms not being reference-cuffed, one by one, to singular inner experiences (for example, a sensation x, or y, or z), but being more diffusely denotative insofar as they are denotative at all, ranging over a broad overlapping, interconnected, intersubjective, comparatively inspectable, quasi-behavioristic, non-privately-secretive, allegedly criterially objective mindscape.

Peter Geach many years ago in a somewhat different but related context made the following remarks about the concept *seeing*:

> To have the concept *seeing* is not even primarily a matter of being able to spot instances of a characteristic repeatedly given in my ('inner-sense') experiences; *no* concept is primarily a recognition capacity. And the exercise of one concept is intertwined with the exercise of others; as with a spider's web, some connexions may be broken with impunity, but if you break enough the whole web collapses—the concept becomes unusable. Just such a collapse happens, I believe, when we try to think of seeing, hearing, pain, emotion, etc. going on independently of a body.
>
> When I apply this sort of concept to a human being, I do so in connexion with a whole lot of other concepts that I apply to human beings and their natural environment. It is easy

> enough to extend the concepts of 'sensuous' experience to creatures fairly like human beings, such as cats, dogs, and horses; when we try to extend them to creatures extremely unlike human beings in their style of life, we feel, if we are wise, great uncertainty—not just uncertainty as to the facts, or as to the possibility of finding them out, but uncertainly as to the *meaning* of saying: "I now know how to tell when an earthworm is angry."[45]

and slightly later:

> Even an earthworm, though, affords some handholds for the application of 'sensuous' psychological concepts; we connect its writhings when injured with our own pain-reactions. But when it comes to an automaton, or again if we are invited to apply the concepts to a supposed disembodied existence, then we may be sure that we are right in refusing to play; too many threads are broken, and the conceptual web has collapsed.[46]

So far as I can tell, Geach is not focusing on the interconnections among cognate psychological concepts as such, but is instead pointing out the relationships of mental concepts to other matters: such as having a body, and so on. But it is his general metaphor of the spider weblike connectedness among concepts that interests me here and which I would like to extend to families of mentalistic notions, and the ways in which such a web might become tattered.

Certainly something like the (no doubt too poetically put) attitudes just mentioned motivated various remarks in Chapter Two. But in afterthought I now find myself vacillating between aspects of the spider web conception expresed there, and the *possible* plausibility of the kind of *Gedankenexperiment* that takes place at the beginning of Condillac's *Treatise on the Sensations*, where he writes under the chapter heading "The First Cognitions of a Man

[45] Peter Geach, *Mental Acts* (London, 1952), pp. 113–14.
[46] *Ibid.*, pp. 114–15.

Limited to the Sense of Smell'' (The typography of Geraldine Carr's 1930 translation has been preserved):

1	The statue limited to the sense of smell can only know odours.	Our statue being limited to the sense of smell its cognitions cannot extend beyond smells. It can no more have ideas of extension, shape or of

anything outside itself, or outside its sensations, than it can have ideas of colour, sound or taste.

2	Only relatively to itself are odours smelled.	If we give the statue a rose to smell, to us it is a statue smelling a rose, to itself it is smell of rose.

The statue therefore will be rose smell, pink smell, jasmine smell, violet smell, according to the flower which stimulated its sense organ. In a word, in regard to itself smells are its modifications or modes. It cannot suppose itself to be anything else, since it is only susceptible to sensations.[47]

The interest I take in the example of Condillac's statue is its implicit suggestion that it makes perfectly good sense to imagine certain mentalistic notions (e.g., rose smell) applying, as it were, solo, to a subject (statue, human being, or machine). For this suggestion at least seems to clash with the spider web construal of the interconnectedness of mental concepts as I am making use of it: namely,

[47] *Condillac's Treatise on the Sensations*, trans. Geraldine Carr (Los Angeles, 1930), p. 3. The *Gedankenexperiment* that occurs at the very outset of Condillac's *Treatise* is in no way indicative of the full-blown theory that he goes on to develop. As his ''experiment'' proceeds, an interconnected network of cognitive-affective capacities quickly emerges. But I am pleading philosophical license in using just the opening ''stanzas'' of the *Treatise* to epitomize an allegedly possible way of imagining how a particular mental capacity might exist in isolation in a subject.

as the view that mentalistic terms apply in families or cognate clusters. (The idea of a statue with the capacity for rose smell, and only that—nothing else in its sensory-cognitive-behavioral repertoire—would, of course, clash as well with the spider web metaphor as Geach is using it. Too many threads would be broken.)

One might, of course, agree on the aptness of the metaphor, yet haggle over proposed examples of ways in which the web would or would not be torn apart by this, that, or the other seemingly vagrant application of a psychological term: 'anger' to an earthworm, 'pain' to a robot, and so on.

Geach's spider web metaphor as I wish to embellish it—and I must emphasize it is a digressive embellishment more than any sort of exegesis—contains two interwoven strands. The first is that mental concepts are, as it were, gossamered together, and that where one is applicable expect many more related ones to apply as well. And, as a kind of corrollary: twiddle around with any one part of the web, and you will set off reverberations in all the other parts. (The spider will sense the fly no matter where it lights.) The second is that by virtue of the interconnectedness of mentalistic concepts and the broad network they comprise, the criteria for the application of psychological terms are spread out, which can make the mind seem, in a sense, "all over the place" insofar as it is any place at all.

Now if one believes that some variation on a spider web metaphor of the mind is apt and could be filled in and made precise, one is also likely to believe that Turing's Imitation Game is an inadequate test for intelligence or thought in a machine. Because endorsement of that test seems to suggest—at least under what I take to be its most plausible interpretation—that satisfying a fairly singular question-answer type of criterion is sufficient for determining the existence of agile cogitation in a subject. Furthermore, since there are no additional claims made to the effect that a machine that passed the Imitation Game test would also thereby exhibit a wide (non-''monorail') array of interconnected cognate notions, one believes that a tacit assumption is being made to the effect that a machine might be able to think, and only think—*whatever* that might mean. And that, certainly, would be at odds with the view of the mind that is catered to by the metaphor of the spider web.

Nevertheless, there is a competing view of the possible non-interconnected solo ascriptions of mental notions to a subject which Condillac's statue, starting out with just the capacity for rose smell, suggests. (Should we so cavalierly dismiss his example as not well imagined?) Here I should mention that it was not actually a browsing of Condillac's *Treatise on the Sensations* that prompted me along this path of thought. Rather, some further problematic reflections on computers brought Condillac's statue back to mind and roughly because of the following considerations. Leave aside, temporarily, high-powered psychological notions such as thought or consciousness, and simply consider computation. At one time computation was viewed as a mentalistic capacity peculiar to human beings and (in the eyes of the gullible) maybe a few talented carnival horses. But now computers can compute—or so it certainly seems. And some computers do simply and primarily that. But how should we account for such a competence if we believe in something like the spider web characterization of the interconnectedness of mental concepts? Is this not more like a Condillac's statue, a mechanized one, to be sure, but one with a solo mental capacity (like rose smell) which happens to be cognitive rather than sensory?

This is the main question I wish to pose in this section, even though what I have to say in response to it is obviously sketchy and quite inconclusive. Nevertheless, here is a sketch of some alternatives:

First, the most simple, and, I think, the most plausible: human computation involves mentality; computer computation does not. So there is no reason to suppose that if computation in human beings has weblike connections with other mental notions, analogues of such connections would be preserved in the case of machine computation. What weak AI teaches us is that the notion of computation is ambiguous with respect to its involving a psychology. Sometimes it does, sometimes it does not.

This view has some ring of truth to it. Yet one needs to circumvent the temptation to treat it as true by definition that "if X-ing is done by a *machine*, X-ing cannot be mental." For, as was argued in Chapter Three, there are problems with such short-shrift approaches to the question whether machines could think, robots feel, and so on. Independent reasons are needed to establish that the possibility of computation by current computers partially depsychologizes the notion of computation, as opposed to its having instead illustrated that computation by machine partially psychologizes machines! and that, further, because it does so, weblike connections among mentalistic notions can be safely viewed as inessential to our understanding of them.

We *could*, on the other hand, revise our view that computation was ever a *bona fide* mentalistic notion after all, and suggest that this is what weak AI has shown us. The reasoning might take the direction of suggesting that although computation when performed by human beings is generally interconnected within a web of other mental notions such as belief, understanding, thought, consciousness, etc., it can be detached from that network as weak AI illustrates, and it is therefore an *essentially* non-mentalistic notion.

Or, even less plausibly, one could take a wildly different tack and claim that computation is indeed mentalistic, *and* it has a spider weblike strand-to-strand interconnectedness with other mental concepts, *and* computers really do not compute. They just appear to. (This latter strategy would be reminiscent of Descartes' redescription of all apparent intelligent animal behaviors.) With respect to this last alternative, I once had a spirited exchange with Virgil Aldrich, who was at the time willing to defend not only the view that computers do *not* compute, but also the view that airplanes do not fly. People, according to Aldrich, compute using computers, and planes are flown by pilots. (Hardly a crazy suggestion!) Obviously one could go down the line and make the same point about specialized types of computational programs such as chess playing, theorem proving, and so on.

And still another (not altogether goofy) alternative might be developed along the lines that although computation in the case of human beings is interwoven with other mental concepts, in the case of computers it is not, and in this respect computation by computers, though mental, differs significantly from computation by human beings. it is, in a sense, "machine indigenous."

Or one could flatly deny the accuracy of the intuitions that prompted the foregoing query: namely, that computers computing, or machines that might be programmed only to play chess, etc., really did lend some credence to a Condillac's statue-type conceptualization of mentality. Instead it could be argued—and with considerable force—that computation, or even a more specialized version of it such as chess playing, is itself a many-splendored thing which subsumes a rich web of interconnected notions: e.g., the capacity for discerning, comparing, evaluating, dismissing, accepting, revising, etc. . . . (It is primarily because of this, one could add, that *computational* approaches to psychology seem at all plausible.)

My own unsettled views on the matter—my primary vacillation—include both some affection for the last-mentioned position, as well as a dose of respect for the view that computation by machine (whether mental *or* non-mental) may be usefully viewed as, in a sense, "machine indigenous" *and in such a way that it operates more like a solo faculty than it does in the case of human beings*; enough so, at any rate, to lend some support to the view that Condillac's statue is after all, in spite of its oddity, well imagined.

This latter perspective, however, I find somewhat at odds with other intuitions I harbor concerning Condillac's statue and the spider web metaphor. Roughly: the class of what I have called program-receptive aspects of the mind seem to me more spider weblike in their relationships than does the class of what I have labeled program-resistant aspects. For example, something like a checker-playing competence—a program-receptive aspect—is more intricately tied to possible intersecting intersocial behavioral patterns—i.e., a significant part of the very criteria for ascribing such a competence is derived from them. Program-resistant mental aspects, however—non-behavioral, non-problem-solving features such as the capacity for feeling a stabbing pain or seeing a blue after-image—seem far more detached from such socially "spread out," "all over the place" conditions for ascription. They seem to be examples of the singular type of sensations—rose smell, et al., that Condillac first bestows upon his imagined statue. But if this contrasting characterization of program-receptive and program-resistant mental aspects is correct, it should seem somewhat surprising that the best parallels between contemporary machine models of the mind and Condillac's statue involve program-receptive features. The surprise, of course, has nothing to do with the fact that machine models exhibit program-receptive features, for most machine models of the mind are program oriented. It has instead to do with the fact that the program-receptive features that are simulated seem instantiated in the machine more in the way that rose smell is imagined as accruing to Condillac's statue—i.e., as a solo capacity—than in the way that, say, human capacities for playing chess are tied in with multiple interwoven weblike connections with other socially shared behavioral patterns. (Note: another thing the surprise I am alluding to does not have to do with is the obvious fact that we cannot at the moment simulate successfully an entire social environment. Some of the issues involved in this will be touched on later.)

Again, my summary thoughts on these problems are transparently inconclusive. But if the foregoing train of metaphor and simile carries any cognitive freight, it might be this: that (1) program-receptive features are (in human beings) more spider weblike in their interconnections than are program-resistant ones, but (2) as they (or their simulations) appear instantiated in current machine models of the mind, they assume more the status of a *solo* capacity—like rose smell for Condillac's statue—which (3) strongly suggests that the way those capacities are incorporated into machines is novel (or "machine indigenous") and hence at best a distorting mirror of human mentality.

The claim that it is often or only as *solo* capacities that human-type mental aspects such as theorem proving, chess playing, and the like have been instantiated in machine models might be dubbed the *idiot savant hypothesis*. It is the hypothesis that insofar as analogues of human mental capacities are represented in machines, they are represented there in highly atypical, isolated ways—like some singular wondrous ability of an *idiot savant*—and not as a competence that, though less efficient and more clumsy, nevertheless radiates out and connects with many other strands of our cognitive-affective existence. (Note: the *idiot savant hypothesis* is not the hypothesis that computer hardware and software is very different from the human brain, which is not in most respects a hypothesis at all but an obvious fact, but rather that the structure of representation of human mental capacities in human brains—conceived of at virtually any interesting level of abstraction—and the structure of representation of whatever their alleged current simulations in machines is, are different—as different as the difference between the calculating wizardry of some *idiot savant* and the arithmetical competence of ordinary human beings.)

In "Lectures on Aesthetics" Wittgenstein is recorded as saying:

> It is not only difficult to describe what appreciation consists in, but impossible. To describe what it consists in we would have to describe the whole environment.[48]

And this point about appreciation, though not exactly the same as, is similar in some interesting respects to the kind of picture conjured up by what I have been calling the spider web metaphor of the interconnectedness of mental concepts. Here, for example, one could pair off Wittgenstein's more diffuse, spread out "all over the place" (as I've put it) view of what appreciation is—not, clearly, some singular phenomenon—with a more monorail (or Condillac's statue) type of approach such as we find in the writings of Tolstoy on art and literature or Clive Bell on visual art. In the theories of both Tolstoy and Bell, the supposition is made that there is a special kind of psychological hydraulics that is triggered into operation, or turned on, whenever a successful transaction between the artist's creation and a reader or viewer takes place. In Tolstoy's account it is the promotion of symmetrical feelings between the artist and the audience that is the earmark of successful art. And in Bell's theory it is a peculiar aesthetic emotion stimulated by what he calls "significant form" in the artifact. In either case there is a discrete type

[48] *Wittgenstein, Lectures and Conversations on Aesthetics, Psychology, and Religious Belief*, ed. Cyril Barrett (Berkeley and Los Angeles, 1972), p. 7.

of phenomenon which constitutes appreciation and which serves as the primary subject matter that aesthetics is supposed to be all about.

Now if one believes that thinking, for example, or feeling, is *more* like appreciation as described by Wittgenstein than it is like a discrete phenomenon, then one could believe that projects in strong AI *human* as well as weak AI *sim.-human* are predicated on false presuppositions. For these projects seem to take it for granted that cognitive processes, like black holes, are a discernibly discrete, though mysterious, subject matter in search of an explanation, which AI in one way, shape, or form may provide. Furthermore, one might also believe that a number of critiques of AI are vitiated insofar as they have the same presuppositions of those projects and simply disagree on how to interpret their tally sheets. Consonant with the conclusions of the position just etched, and utilizing Wittgensteinian strategies in original ways together with arguments wholly his own, Michael Reed in his "Cognition and Explanation: Artificial Intelligence and the Philosophy of Mind"[49] has provided a forceful, though bound to be controversial, critique of both AI and various of its detractors. With respect to Newell and Simon's approach to AI, Reed writes of their treatment of "general intelligence" as a "putative "phenomenon"" as follows:

> We have seen how the picture of empirical inquiry which they offer determines the picture they must have of intelligence, *qua* scientific subject-matter. It must be a "phenomenon" in "Nature," specifiable in general terms, so as to be properly

[49] Unpublished Ph.D. Dissertation, University of Minnesota, 1984. Reed has his own way of formulating the problems facing AI and its critics, and of arriving at the conclusions I am suggesting are parallel to the above-mentioned ones. So I do not wish to pin any of my metaphors onto his phrasings, nor suggest that he would be at all happy with how I have delineated or dealt with any of the issues contained in this section.

identified for scientific investigation. If this picture were to be applicable to the intelligent human actions we see "everywhere around us . . . in human behavior," "intelligent action" and related expressions would have to be put to use in those everyday surroundings in the same way as are the subject-matter expressions in empirical inquiry as Newell and Simon understand it.

But we also saw that in the very situations which Newell and Simon say exhibit the subject-matter of AI, there are no such "phenomena" to be found. Their way of characterizing intelligence and "intelligent action" as "phenomena" in "Nature" does not fit the facts as they actually exist in those situations. The very same negative result was arrived at in our discussion of the allegedly relevant area of computer system and design. The actual roles played by the notion of 'intelligence' and related notions in that field display no discernible relation to any "phenomenon" called "general intelligence." There is, *a fortiori*, no support for the claim, argued for by Newell and Simon and widely subscribed to by others, that what goes on there provided the means for explaining what Newell and Simon want to refer to as "intelligent action."[50]

Although I cannot here address myself to any of the details of Reed's critique, I believe at least parts of his analysis could be used to support a spider web-type depiction of mentality, as opposed to a Condillac's statue rendition. And, at the very least, the above quotations should help illustrate the possibly high methodological stakes involved in selecting one portraiture rather than the other.

My own restless view of these matters currently is that program-resistant aspects of the mind are more like Condillac's statue's rose smell than they are "all over the place." And that because of their manner of being entwined with program-receptive features of the mind—no problem solving without some consciousness of it, for example—they give those features more of a locus as

[50] *Ibid.*, p. 79.

well—gather them into a discrete center of awareness, and keep them from being in fact as spread out as the admittedly diffuse and multiple criteria for our ascribing them to others would suggest. But all this, admittedly, is not so much an argument on behalf of a hypothesis, but rather, a hypothesis in need of many arguments.

Further Problems and Emotions Concerning Program-Receptive and Program-Resistant Aspects of the Mind

It should be plain from the foregoing chapters that the distinction between what I have called program-receptive and program-resistant aspects of the mind is a wholly general one, and not dependent on any particular program or set of programs. That it was illustrated in connection with IPL-V (in Chapter Five) is incidental and in a sense superfluous. For the distinction is preserved across all types of programmed computers.

Two primary points it was designed to accentuate can be rephrased in the following way: (1) whatever the human mind is, it is more than a program, and much of the more than a program that it is provides constraints on the kind of a program it can contain; (2) it is a categorial mistake to attempt to model all aspects of the mind through programming.

Notice: the first point is different from the point made by Searle that you cannot get semantics from syntax, and from the contention that the mind is more than mere computational processes. It is rather that even if you add to syntax or computational processes a dimension of semantics (or intentionality), the mind is also more than that. And the more than that that it is involves such features as basic input potentials for pain, anxiety, elation, afterimages, and so on. And the presence or absence of any or all of these severely limits the type of programming that is possible.

Here is should be pointed out that although I tend to associate program-receptiveness with software and program-resistance with that which is being programmed, hardware (or brainware), the distinction does not depend on a hard and fast distinction between softwares and hardwares.[51] For example, what is represented in one computer system's pseudo-code (form of software) may be present only in another system's machine language. And so on. That is, the distinction between software and hardware has over the years been to some extent a "floating one," with some sorts of programming—say, in a so-called machine language—being viewed as more hardwarelike, closer to the machine, than, for example, more abstract levels of programming that at some point need to be translated back into machine language. So to describe a computer's capacities at the level of "machine language" *could* be viewed as both a programming and quasi-hardware level of description. Nevertheless, it seems to me that machine-language programs are still clearly programs whether or not they are viewed as the manipulation of hardware. And even if an ambiguity exists as to whether programming at the level of a machine language should be seen as hardwarelike, it has no bearing on the basic program-receptive/program-resistant distinction. For, all that distinction needs in order to be viable is the distinction between a program and whatever other than a program is required for it to be operative within a given system (or subject).[52] There could not be, for example, a

[51]Obviously both program-receptive and program-resistant aspects of the mind depend in turn on other *non*-mental features such as whatever is necessary for embodiment or some counterpart thereof. I confess to a general neglect of the issues involved in this and have been roundly thumped for it and other omissions in an interesting work by William John Holly entitled "Program-Resistant Aspects of Knowing-That and Knowing-How", unpublished Ph.D. Dissertation, University of California, Irvine, 1975. I hope to deal with some of these criticisms later.

computer system that was solely describable in terms of a program.[53] (Compare: arguments against the intelligibility of disembodied spirits or minds.)

So if it makes sense, as it surely does, to talk about programming (which I *generally* associate with software), it also makes sense to talk about *there being something* that receives the program. And that is what I am referring to as hardware or brainware. Such a general dichotomy is all that is needed to give a conceptual base to the distinction between program-resistant and program-receptive aspects of the mind. In a computer-independent way, the distinction can be described as one between the physical input capacities, whch are prerequisite to the having of certain mental states or events—pains, fears, anxieties, afterimages—and the abilities (super- and subroutines) necessary to performing various tasks, executing plans, and carrying out intentions, and the like.

The significance of this is not simply that hardware (or brainware) limits software, but that hardware (or brainware) *under certain interpretations*, viewed as types of basic experience potentials, such as for having a pain, feeling an anxiety, having an afterimage, etc., limits programmability and is not itself analyzable in terms of it. Looked at in this light, it is apparent that there is an important sense in which the program-receptive/program-resistant distinction could only be incompletely reflected by current hardware/software distinctions. For *both* program-resistant and program-receptive aspects of mind demand double characterizations: from third-person and first-person points of view.

[52] This should, I think, mitigate the type of criticism leveled against me by James H. Moor in his "Three Myths of Computer Science," in *British Journal of the Philosophy of Science*, 29 (1978), pp. 213–22; especially pp. 215–16.

[53] And to think that there could, might fairly be labeled a modern form of Cartesianism. See Searle in Haugeland, *op. cit.* p. 304.

There is a problem, however, that pops up like a macabre Jack-in-the-box when trying to make this point explicit, a problem which is itself interesting. In the case of human beings, what I have called program-resistant, non-problem-solving, non-behavioral aspects of mind have been exemplified by such things as the basic capacities for pain, afterimages, anxieties, etc., which once instantiated in *some* (as yet unexplained) analogous way in a machine would make certain problem-solving quasi-behavioral features of the mind receptive to programming. But comprehension of a program-resistant (basic capacity) feature requries it to be viewed as both (1) a basic capacity (such as the capacity of a computer to print its output in blue ink) *and* (2) a mental feature (something involving an immediately known conscious experience). That seems fine, except that we are not thereby provided with any way of seeing how the essentially third-person, basic-capacity component of a program-resistant aspect of the mind is connected with, or the same as (or what!?) its first-person psychological component. In other words, as I will outline later, some nasty problems attend any attempt to provide a double-barreled first- and third-person characterization of program-resistance and program receptiveness in a nontrivial, non-question-begging way; a way that helps characterize the mind without reiterating the various perplexities that motivate the characterization.

It now seems clear to me that if the program-receptive/program-resistant distinction is viable—as I still believe it is—its most *unproblematic* utility lies in a certain taxonomic suggestiveness, and as a kind of diagnostic distinction that can help cross-hair different types of problems in mechanistic modeling. And that was, of course, the major motivation for cooking it up in the first place. Nevertheless, it should be emphasized that as currently formulated it sheds little light on issues such as the mind-body problem and the problem of other minds, although, hopefully, it could contribute to an alternative way of rephrasing these problems.

Before saying anything further about the just-touched-on inscrutable matters, however, it seems useful to add a refinement to the program-receptive/program-resistant distinction. Obviously there is a miscellany of items wrapped up in our mental life which includes our ambitions, plans, wants, desires, commitments, engagements, and so forth, which we could not now program a robot or machine to literally partake of (strong AI *sim.-human*). This would include such acts as signing legal contracts, getting baptized, marrying, running for mayor of Minneapolis, and so on. But the reasons it is not now possible to program a robot or machine to be baptized, get married, or run for mayor are quite unlike the reasons we cannot *program* a robot to feel pain. And this is because providing a robot with the capacity for pain *is not* a problem in programming! In contrast, getting married, for example, is like typical program-receptive features of the mind, associated with well-defined tasks, specific achievements, successes and failures, and is also "protocol possible" in that one could provide verbal reports on what one was doing in getting married. The unprogrammability of the actions requisite to baptism, marriage, and a myriad of other complicated socio-cultural-religious-legal actions is essentially institutional, whereas the problem of providing a machine with the potential for pain has to do with problems in embellishing the robot's or machine's basic underlying capacities.

This added "cultural" or institutional aspect of programmability is, I think, consonant with points developed by Dreyfus in his "From Micro-Worlds to Knowledge Representation: AI at an Impasse" (1979, 1981). Dreyfus discusses Winograd's program SHRDLU for natural language comprehension, which, as previously mentioned, involves simulation of a robot arm that moves blocks around and enters into question-answer and command conversations about its blocky universe. In connection with a conversation with SHRDLU involving owning, Dreyfus quotes some remarks by H. Simon (1977), which include:

> "SHRDLU's test of whether something is owned is simply whether it is tagged 'owned'. There is no intentional test of ownership hence SHRDLU knows what it owns, but doesn't understand what it is to own something."[54]

Dreyfus goes on to claim that SHRDLU "couldn't own anything, since it isn't part of the community in which owning makes sense. Given our cultural practices which constitutue owning, SHRDLU cannot own something any more than a table can."[55]

Dreyfus's claims about SHRDLU and owning seem to me apt and an excellent example of the type of point that can be made about the non-programmability of computers or robots to perform various acts typically carried out within an environment of finely tuned cultural-institutional standards of permissiblity and prohibition. But what, then, is the relation between *this* type of non-programmability, as I prefer to call it, and what I have labeled program-resistant aspects of the mind?

At first it might seem that the category of program-resistant mental aspects could simply be divided into two kinds: (1) those having to do with basic capacities or input potentials for pains, emotions, and other sorts of non-problem-solving, non-behaviorial features of conscious experience, and (2) those having to do with various sorts of actions, doings, and what-not, the possibilities of which are severely constrained by cultural-institutional factors. Yet I think there are some strong reasons for resisting this subdivision within the category of program-resistant mental

[54] Dreyfus, in Haugeland, *op. cit.*, p. 169.
[55] *Loc. cit.*

features. (Here I am simply assuming *contra* various extreme approaches of the neo-Wittgensteinian type that there is a significant hard-core competence to (at least) our program-resistant conscious life which includes pains, sensations (and even certain emotions) which we experience quite independently of *any* cultural-institutional, socially constrained sorts of factors.)

Recall that (as pointed out in Chapters Three and Five) program-resistant aspects of the mind are features which it does not really make sense to attempt to construe in terms of well-defined tasks, problem-solving types of behaviors, and so on. But that does not seem to be true of such things as getting baptised or married, running for political office, and so on. Although it is not *now* possible for a robot or machine to be programmed to attempt to execute the kinds of plans (routines, subroutines) associated with the last list of tasks, that does not mean that it does not make sense to imagine someday doing so. *That* does make sense, and what one has to imagine in imagining that is a vast change in the cultural-institutional context in which such machines or robots would "live," move, calculate, and produce their outputs. But what this, then, strongly suggests is that currently non-programmable features of our lives such as ownership, baptism, marriage, running for political office, and the like, are not only *not* a subclass of program-resistant mental aspects, but a subclass of program-receptive ones. Support for this interpretation can be gleaned from the fact that if we consider cultural-institutional acts such as getting married, et al., from the standpoint of a project in weak AI *sim.-human*, it at least *makes sense* to attempt to devise a program that simulates such an act, in the same way that it *makes sense* to attempt to devise a program that will enable a machine to simulate the act of criticizing poetry with the dexterity of a reviewer for *Time*. (A program for simulating getting married would no doubt contain some micro-representation of *contexts*, within which getting married now occurs, and so on.) Obviously, given the current state of the art—and, perhaps, *any* future state of the art!—neither of the two imagined weak AI *sim.-human* programs would be very convincing or interesting as models of human socio-cognitive performances. And that is in itself interesting and important. But it is not the point I am concerned with here, which is simply that it *makes sense* to provide a representation within programming terms of the acts of getting married or criticizing poetry with the dexterity of a reviewer for *Time*. Furthermore, it must be emphasized, the *kind* of representation that makes sense is not just a kind of *by fiat only* representation, where one, as it were, simply says—or almost simply says—"Let *this*" (singling out some aspect of software) "represent *that*" (a human feeling or emotion, for example). It is rather that various nonarbitrary structural mappings can take place between, for example, the elements of a program and human protocols or other representations of human behaviors, such as in the work of Simon, Shaw, and Newell as discussed in Chapter Four. (*By fiat* representation is touched on in the last section.)

That an aspect of our mental life is program-receptive tells us nothing about whether we can now or ever devise a theoretically interesting or illuminating program either in the sense of strong AI *sim.-human*, or of weak AI *sim.-human*. Compare: we can know that if we construct certain patterns of words, we will write double quatrains, but we do not thereby know if they will be good, bad, or indifferent poems. In other words, the notions of program-receptive and program-resistant aspects of the mind are exceedingly general and evaluation-neutral. But, we hope, they have a role in helping triangulate those features of our lives that could, in principle, be subjected to mechanistic modeling through programming, and thereby call into relief those that could not.

To highlight the distinction between what *at a given time may be unprogrammable* and what I have called program-resistant aspects of the mind, consider some currently intractable pattern-recognition problems. A number of these have been discussed by Dreyfus,[56] et al., and they include such matters as trying to conceptualize in programming terms what it would take to endow a machine with the capacity for recognizing items having certain uses or functions such as archways, chairs, and so on. No doubt whatever is involved in human recognition of such things, it amounts to more than simple perceptual sensitivity to exhibited "surface"[57] properties of the objects, and must include, somehow, an understanding or awareness of ways in which we treat or use them. Such objects, omni-

[56] Dreyfus, in Haugeland, *op. cit.*, pp. 161–204.

[57] A distinction popular within aesthetics in recent years between "exhibited" and "non-exhibited" characteristics or properties might prove useful here. George Dickie in his influential "What Is Art?: An Institutional Analysis" writes:

> Recently, Maurice Mandelbaum has raised a question about Wittgenstein's famous contention that "game" cannot be defined and Weitz's thesis about "art." His challenge to both is based on the charge that they have only been concerned with what Mandelbaum calls "exhibited" characteristics and that consequently each has failed to take account of the nonexhibited, relational aspects of games and art. By "exhibited" characteristics Mandelbaum means easily perceived properties such as the fact that a ball is used in a certain kind of game, that a painting has a triangular composition, that an area in a painting is red, or that the plot of a tragedy contains a reversal of fortune. Mandelbaum concludes that when we consider the nonexhibited properties of games, we see that they have in common "the potentiality of . . . [an] . . . absorbing non-practical interest to either participants or Spectators." Mandelbaum may or may not be right about "game," but what interests me is the application of his suggestion about nonexhibited properties to the discussion of the definition of art. Although he does not attempt a definition of "art," Mandelbaum does suggest that feature(s) common to all works of art may perhaps be discovered that will be

present in our experience, are created by ourselves with various purposes in mind, and an understanding of these seems presupposed in our ongoing capacity to recognize them for what they are. Our perception of objects with uses or functions seems to involve multiple dimensions of sensory-cognitive awareness which may not be operative in, for example, the recognition of more "natural" Lockean-type substances such as cats, dogs, gold, water, and elm trees. Furthermore, human pattern-recognition competencies include the ability to detect in our various modes of behavior such things as obligations, punishments, promises, and praise, or what Locke called "mixed modes." Such "non-things," according to Locke,[58] in-

---

a basis for the definition of "art," if the nonexhibited features of art are attended to.

The above passage originally appeared in Dickie's *Art and the Aesthetic—An Institutional Analysis* (Ithaca, 1974) and was reprinted in an anthology edited by Dickie entitled *Art and Philosophy*, 2nd ed. (New York, 1979) pp. 83–84.

If there are, as it seems to me undoubtedly there are, crucial nonexhibited properties of things *and* non-things which we, however, with ease and alacrity are able to detect, discriminate, and discuss day after day within the framework of our garden-variety experience, the question arises as to how such items can be accommodated by pattern-recognition theories. What is it that accounts for our awareness of nonexhibited properties and supplements our perceptual capacities which are otherwise most often tuned primarily to detecting shape and topological characteristics?

[58] The phrase "non-things" is not Locke's, nor is it an altogether accurate way of rendering what falls into his category of mixed modes. Relations, which could also be viewed as non-things, are not treated as mixed modes by Locke. It should also be pointed out that mixed modes themselves constitute a mixed (heterogeneous) class that includes fights, adultery, obligation, punishment, and so on. (Most of our ethical and aesthetic terms would, according to Locke, be *names* of mixed modes.) See *An Essay Concerning Human Understanding*, by John Locke, Vols. I and II, ed. A. C Fraser (New York, 1959), especially Bk. II, Chap. XXII (Vol. I) and Bk. III, Chap. V (Vol. II).

volve, for the most part, distinctions we design freely and creatively for our own convenience. And with respect to a broad subclass of these mixed modes, we do not, according to Locke, simply stumble upon them in nature as we might stumble upon gold or dung beetles or elm trees. Again, as with functional objects, our sensory-cognitive awareness of mixed modes appears to involve competencies not nearly as easily conceptualized, much less now realizable in a machine, as, say, the ability to discern that something is gold or a dung beetle or an elm tree. Given the current state of the art, it seems that there is no pattern-recognition program that captures in an interesting manner our ability to recognize functional or mixed-mode features of our experiences. But again, this is a point about what *at a given time* is unprogrammable. It does not serve to show that such kinds of perceptual-cognitive abilities are program-resistant. They are, I think, in spite of the confusions they generate, program-receptive. And to try to imagine how analogues to them could be incorporated into a program might be of some help in clarifying this opaque feature of our minds. Here one takes what one can get. It is not as if other (one or more) non-machine-oriented explanations already exist that are so convincing we need only to barter about the details.

Some Brief Comments on the Connection between Cognitive and Moral Decisions Concerning Robots and Machines

Whether or not a given subject is capable of planning, intending, and/or performing certain kinds of acts depends not only on who or what they are, but on how they are treated by others, whether or not those others are like or unlike them in some, most, or all respects. It is possible for me to plan to and actually apply for and receive a home repair loan from a credit union because of how others treat someone such as myself with a certain income and credit rating. My ten-year-old son, however, could not take out a loan for anything from a credit union. Or consider horses. They are made capable of winning things, being honored, and so on, because of the institution of horse racing. On the other hand, ants are not candidates for the sort of praise that would accompany winning the Triple Crown. Young children are incapable of getting married because of various laws, customs, and conventions. And then, of course, there are the sadly familiar constraints and prohibitions on kinds of plans and behaviors that are due to prejudice about race, sex, age, and so forth. And these deserve special mention. Let me call this latter set of constraints *prejudicial constraints* in order to single them out from other cultural-institutional constraints on a given subject's plans, intentions, and behaviors. I do not mean to analyze these here—though they are obviously deserving of concerted philosophical analysis. All I wish to do is separate them from what I would regard as essentially reasonable (or harmlessly whimsical) cultural institutional factors, which either make possible or prohibit various plans, intentions, and performances: the prohibition on loans by credit unions to ten year olds, the possibility of a horse winning the Triple Crown, the impossibility of an ant doing something parallel to that, and so on.

Regarding prejudicial constraints, there is, perhaps, no reason to suppose that the situation would (will?) be any different for us vis-à-vis our artifacts should they ever attain to anything like the cognitive-affective status envisioned for them by the more hopeful proponents of strong AI *human*.

A rather optimistic prediction, however, with respect to the hurdle presented by cultural-institutional constraints on mimicking human beings with machines is given by an attorney, Marshall S. Willick. In his article "Artificial

Intelligence: Some Legal Approaches and Implications'' (1983), Willick writes:

> The history of law in the United States is punctuated by the extension of legal recognition and rights to an increasing number and variety of groups. The precise rationale for each extension varied considerably, but each represented an acknowledgement that the individuals comprising the group being considered were more like the persons doing the considering than like the property belonging to those persons.
>
> As computers behave increasingly like humans, the reasonableness of treating them as persons will increase. This is so because treating an individual that *appears* to be a person as more property calls the validity of the distinction into question and thereby weakens the foundation of society.
>
> Neither the total mechanization of the human body nor the computerization of the brain yields a point at which a person should devolve to a mere machine. If a mechanized body and brain may be treated as persons in *some* instances, why not in all?
>
> Eventually, an intelligent computer will end up before the courts. Computers will be acknowledged as persons in the interest of maintaining justice in a society of equals under the law. We should not be afraid that that day may come soon.[59]

Related to Willick's remark are those made by Aaron Sloman and Monica Croucher in their article ''Why Robots Will Have Emotions'' (1981):

> Many people deny that machines could ever be said to have their own motives. Machines hitherto familiar to us either are not goal-directed at all (clocks, etc.) or else, like current game-playing computer programs, have a simple hierarchical set of goals, with the highest-level goal put there by a programmer. If machines were designed with a system of motives and motive generators as complex as that described above, then the machine could develop and change over time in such a way that it would be misleading to say that the machine was

[59] In the *AI Magazine*, Summer 1983, p. 14.

pursuing the goals of its designer. Ultimately the decision whether to say such machines have motives is a *moral* decision, concerned with how we ought to treat them.[60]

Although Willick seems to me to be jumping the gavel, his article raises interesting issues about cultural-institutional constraints across species and implicitly, at least, suggests how this influences ascriptions of mentality to machines, which in turn bears on the question of artifactual rights (my phrasing, not his). Furthermore, I think that he is correct to treat the issue of extending rights to machines as being in some ways a function of the extent to which they appear to be persons. I want to add, however, that whether computers or robots appear to be persons will (or definitely should!) depend on more than descriptive similarities involving behaviors—much less parallels at the end-results level of comparison. The various inadequacies of Turing's Test underscores that. So, too, there may be *less* than similarities in behaviors involved, and a robot or machine could be deserving of rights even though in many ways it failed to perform in a range of ways comparable to human beings. What qualifies a robot to be treated as a robson or persot (recall Chapter Three) will involve not only its program-receptive abilities to execute tasks, but its underlying program-resistant repertoire as well, together with a medley of biographical and other historical and ecological factors. The issue of extending rights such as we grant ourselves to robots or machines and the myriad of hybrids in between them and persons will (should) depend not only on how similar to us they are in what they do, but on how similar they are to us in what they are.

Sloman and Croucher are correct in sensing a hookup between the sociological facts about how we treat ma-

[60] In Ninth International Joint Conference on Artificial Intelligence, Proceedings, 1981, p. 200.

chines, and what we can ascribe to them, but I think they mischaracterize that connection.

Whether or not we should ascribe motives to machines seems to me a question mixed up with a blend of factual and, to a lesser extent, conceptual considerations. In this respect motives are like anything else: vision, emotion, belief, pain, and so on. Furthermore, *and for the most part*, it is how our beliefs involving these factual and conceptual issues get divvied up that in turn affects how we end up treating machines, and not vice versa. In other words, I assume we do not punish or reward machines or robots in the way we punish or reward people, and that this is because *we know* (or *believe* we know) that they lack various basic capacities to begin with. So too, at this juncture, I would assume machines or robots do not have rights because the granting of rights makes no sense prior to our being convinced that they are endowed with certain basic capacities. If machines are viewed as lacking certain basic capacities and viewed as bereft of desires, needs, wants, and so on, we are unlikely to ascribe anything like motives to them, and accordingly will resist treating them in certain ways. It is not the reverse: that we will find ourselves attributing motives to them *only* after we begin treating them in certain ways.

To ascribe or withhold ascription of motives to robots or machines at this stage of their development is not anything like a moral decision, as Sloman and Croucher suggest. It is simply a good old-fashioned, complicated, psycho-philosophical, factual-conceptual matter. Moral considerations arise only after those more basic ontological matters seem more or less settled.[61] That is, the question how to treat robots or machines makes sense only after we understand what sorts of "beings" they are. Suppose, for example, in our collective cybernetic wisdom we decided that robots or machines did, indeed, have certain basic capacities, were the *self*-perpetrators of various complicated acts, and in general satisfied the research goals of "strong" AI *human*. And, further, suppose that we nevertheless refused to treat them in any way like persons. Then, certainly, it would be accurate to say prejudicial cultural-institutional (personal) constraints blocked the ascription of significant mental attributes and behaviors to those robots or machines. At this point what we are willing to ascribe to them becomes a function of how we are willing to treat them, and moral decisions of the kind suggested by Sloman and Croucher take on an urgency.

It is important to see, though, that such moral decisions are not thereby playing anything like a comparable role all along the line as technology develops and machines and robots become more sophisticated. Indeed, nothing could count as a moral decision for us in connection with a machine or robot prior to our already having decided on factual-conceptual grounds that such a robot or machine had capacities sufficient for us to treat it in certain ways. Compare: deciding whether a thermostat or a flower or an ant or a penguin can feel pain, think thoughts, form motives, lay plans, and so on, is not a moral decision about how to treat them. It is a cognitive decision about the kind of creatures they are. But once having made a cognitive decision concerning their natures, the possibility or impos-

[61]There are admittedly troublesome exceptions to this point such as cases in which, because of prejudicial attitudes to begin with, conceptual-factual ontological judgments are affected. For example, because one owns slaves, their *natures* are described as very different from that of their owners; because one loves to eat fish, fish are described as without *any* cognition or emotion, and so on. But in spite of the possibility of various species-chauvinist judgments along these lines, I do not think that in any significant way *so far*, they have impinged on our ontological assessments of robots, and believe that to date it is because of relatively non-prejudicial ontological assessments that the ascription of motives and the like to them has been withheld.

sibility of making moral judgments about them will become clarified.

The Impasse as Regards Mechanistic Modeling, Program-Resistant Aspects of the Mind, Consciousness, and the Mind-Body Problem

Although hardly a fresh insight, it was previously remarked that to model a reaction to having a pain in one's toe (emitting "ouch" for example) is not the same as modeling a pain in the toe. For the so-called qualia—the experience of the pain, the hurt—would not thereby be represented. To have said this, of course, is not, unfortunately, to have contributed much to an understanding of what a model of consciousness or "qualia" might be like. Sometimes I tend to think such understanding can never be forthcoming, and that the very notion of modeling is hopelessly harnessed to third-person perspectives, whereas consciousness or "qualia" is an essentially first-person one. In other words, to be able to sketch what a model that incorporated a representation of consciousness would be like is equivalent to being able to sketch a model of subjectivity. And this we do not know how to do, which may be, in the last analysis, the same sort of ignorance as our ignorance of the nature of the mind-body relationship itself. This, I take it, was the point of the robot's remarks quoted at the beginning of this postscript. As Thomas Nagel put it:

> The idea of how a mental and physical term might refer to the same thing is lacking, and the usual analogies with theoretical identification in other fields fail to supply it. They fail because if we construe the reference of mental terms to physical events on the usual model, we either get a reappearance of separate subjective events as the effects through which mental reference to physical events is secured, or else we get a false

account of how mental terms refer (for example, a causal behaviorist one).[62]

The crucial matter is this: to aim at (now, and maybe always) an actual incorporation in a strong AI project of program-resistant aspects of mind and the phenomenology they involve seems hopelessly utopian. But any of the more modest alternative models one can imagine seem haplessly prosaic. They all have that *representation-by-fiat-only* ring to them: as if by simply naming some operation or item in a program or flow chart and saying "*This* (affect, etc.) is represented in the program by *that*" and then proceeding merrily to describe input-output patterns, one believes that an illuminating model of affect has taken place. This is all too much like drawing the pineal gland, making a dot in it, and saying "This [pointing to the dot] represents the human soul as it is joined to the human body." Well of course it does, but that tells us nothing instructive about the human mind and its relation to the body, but only something about how easygoing and unilluminating certain forms of representation can be. Herbert Simon's (1969) representations of motive and emotion in his "Motivational and Emotional Controls of Cognition"[63] seemed to me unilluminating in this way. And Colby's (1981) "*conceptual representation*" of feelings of shame in his model of a paranoid mind (mentioned earlier) impresses me as equally unconvincing.

In *Brainstorms* (Chapter 2, "Reply to Arbib and Gunderson") Daniel Dennett wrote:

If we are to capture the program-resistant features in an artificial system, we must somehow give the system a phenomenology, an inner life. This will require giving the system something about which it is incorrigible, for whatever else

[62] In "What Is It Like to be a Bat?" in *Mortal Questions* (Cambridge, 1979) p. 177.

[63] See Chapter Five, Section IV.

> one must be to have a phenomenology one must be the ultimate authority with regard to its contents.[64]

Dennett's first claim in the above quotation strikes me as correct and consonant with my foregoing remarks. But I think he mischaracterizes what giving a system a phenomenology or inner life involves, when he links that to incorrigibility. (It must be emphasized that he no longer does this, though, I suspect, the idea that incorrigibility is a mark of the mental still persists in the minds of some.) Dennett is accurate in construing an inner phenomenological life—our conscious experiences—in terms of that to which each person has a privileged access, and, I would add, to which any person other than ourselves has at best an underprivileged indirect (inferential) access, or no access at all. But it should be underscored that it does not follow from our having privileged access to our experiences that we could not be mistaken in our reporting on them (that they would be incorrigible). Rather, there is a way in which we are aware of them and can report on them (whether accurately *or* inaccurately!) that cannot be duplicated by anyone else with respect to those same particular experiences. No one else can have the same relation to the data-base for *my* pain reports as I do—namely, my having of those pains. So the problem arises as to how to incorporate the analogue of conscious experiences and the special privileged access we have to them into a machine model in some non-trivial (''non-representation-by-fiat'') manner.

Notice, a kind of irony surfaces from all this when thought of in the following fashion: that it was partly because of the theoretically unmanageable features of consciousness in the first place that machine models were devised. A number of philosophers—at one time myself included—used to think that a possible way to circum-

[64] (Montgomery, Vermont, 1978), p. 32.

scribe the problems that consciousness raises for the mind-body relationship and the problem of other minds was to try to conceptualize how a mechanistic model of mentality might be constructed in some non-vitalistic, non-emergent, ghostless, non-Cartesian, step-by-step fashion. But even though machine models of the mind have proliferated and evolved in sophistication, they have in each and every case exhibited a glaring lack of any nontrivial representation or instantiation of an analogue of "qualia" or conscious experiences or what is closely entwined with program-resistant aspects of the mind. So, too, since program-receptive aspects of the mind are entwined with program-resistant ones—human calculations, for example, are generally accompanied by conscious awareness—that lack is conspicuous with respect to the modeling of program-receptive features as well. In other words, the problem of providing an acceptable depiction of the connection between consciousness and the physical, whether the physical be neurological or cybernetic, has found reiteration within the mechanistic models themselves, suggesting that the efficacy of such models in helping us to comprehend the mind-body relationship may be nil.

In a more comprehensive fashion the point may be put as follows: no model (and hence no machine model) of the human mind could be an adequate model unless somehow incorporated within it was a non-trivial (i.e., not by *fiat*) representation or analogue of whatever the psychological conditions are which at least make it appear to philosophers (and others) that there is a mind-body problem (as well as a problem of other minds). In other words, an exceedingly interesting and important fact about human minds is that they are capable of being baffled about themselves in certain ways! This is to be sure an *exquisite* fact about them, but a fact nevertheless. Let me label the need to represent this fact in any adequate model of the human mind the *mind-body (and other minds) perplexity condi-*

tion, or, for short, *the perplexity condition*. Furthermore, any adequate model of the human mind not only must satisfy *the perplexity condition*, but must satisfy it in such a way that it does not somehow incriminate the model itself—i.e., by suggesting that the model is as opaque as the human mind seemed to be prior to the existence of the model. This might be called *the transparency condition* of adequacy for modeling (and hence machine modeling) of the human mind. All this, in a sense, is just a much more elaborate way of saying that any ultimate success at machine modeling of the mind must be able to counter the view anticipated in the Epilogue to the effect "that even if we were to construct machines that could think or robots that had feelings, nothing philosophical would be advanced thereby, for all the traditional mind-body problems would then arise for machines and robots."

Perhaps some of the pessimism reflected in the above needs to be muted. In connection with this issue, K. V. Wilkes made the following suggestion in her "Functionalism, Psychology, and the Philosophy of Mind":

> Above all, consciousness *per se* is not a single phenomenon which as such provides a manifestly self-evident *explanadum* for any adequate psychology. It may even prove to be best understood as a second-order property, like intelligence: i.e. a property that becomes ascribable once a sufficient subset of first-order psychological predicates can be ascribed. If so, it would be a function of the complexity and flexibility of the system in question.
>
> The most powerful current attempt to pick out what is special about the phenomenal is the use of Nagel's question, "What is it like to be an X?" The suggestion is that there is nothing that it is like to be a robot—*whatever* its S-F design—whereas there is something that it is like to be a bat or a human. Insofar as I follow this objection, it seems to me to beg the question in advance: certainly there is nothing that it is like to be any contemporary IBM computer, but all com-

> puters so far devised are relatively incredibly simple (a recent notional IQ test on the most sophisticated computers brought them out on a level with the earwig). What we would say if a system was constructed which had, throughout the range of human capacity, substantial functional and structural isomorphism with a human agent is surely a question that nobody could possibly answer a priori at this stage.[65]

And maybe she is right—that we cannot answer it *a priori* at this stage. But there are, I think, some plausible things leading up to an answer that we can say, things which strongly suggest that we cannot now even conceptualize how any *present or future* machine models of the mind would help untie Schopenhauer's "world-knot," the entanglement of the mental with the physical. What I find unclear in Wilkes's suggestions is not so much what it would be like to decide, on the basis of some (any!) reasonable criteria of complexity and flexibility, that a given system had achieved a sufficient degree of it so that consciousness had at last become ascribed to it, but how even if that should come about, consciousness would thereby be exhibited in some illuminating way.

This unclarity is not idiosyncratic, however, but seems democratically distributed. It attends Dennett's suggestions in *Content and Consciousness*, mentioned earlier, that there could be a system of internal states or events, the extensional description of which could be upgraded into an intentional description. We should first note that by couching the problem in terms of how we could upgrade some physical system to a status where ascriptions of intentionality would seem reasonable, we (inadvertently) shift the mind-body problem away from any consideration of first-person psychological perspectives. This becomes clear if we try to ask ourselves the above-mentioned question of

[65] In *Philosophical Topics* (formerly *Southwestern Journal of Philosophy*) XII (1981), p. 166.

Dennett's. To do so we must first presume *we* are some sort of physical system, and then ask, with some punch to the probe, whether we might upgrade ourselves to the point where intentional descriptions would seem applicable. There are at least two things wrong with this rendition of the question: the first is that it would seem to presuppose that there is no difficulty in viewing ourselves as a purely physical system; the second is that it seems to presuppose that we do not already know that we are knowing subjects to which intentional descriptions are applicable. (Compare the oddity of asking: Am I the sort of creature that can ask questions? come up with answers? and so on.)

But where the way in which I can raise the question of intentionality about one thing (namely, myself) is altogether different from the way in which I can raise the question about something else (a neurocybernetic system, for example), there is a reason for thinking the former thing is in crucial respects radically different from the latter and could not be a plausible model of it. Whether or not by some trick of logic intentional sentences could be ontologically defused by finding extensional equivalents, or whether, more plausibly, physical systems might, *through* Dennett's tactics, be "upgraded" into candidates for content ascriptions, the following problems would be left dangling for the class of intentional sentences. Consider any intentional sentence of the form "I ________ that there are gophers in Minnesota" where "________" is to be filled in by any intentional verb (*believe, suppose, think*, etc.), and contrast our way of knowing its truth (or falsity) with any non-first-person variant thereof such as "Dennett (he, she, you, they, it)________(s) that there are gophers in Minnesota." Even if such intentional sentences could be paraphrased by some purely extensional equivalents, this would do nothing to undercut the fact that a sharp cleavage exists between the way in which we know the truth (or falsity) of first-person versions of such sentences and the way we know the truth (or falsity) of their non-first-person counterparts. That is, if I know that "I suppose that there are gophers in Minnesota" is true, the way in which I come to know it is radically different from the way I might come to know that "Dennett supposes that there are gophers in Minnesota" is true. In my own case I know through having supposed that there are gophers in Minnesota, whereas I can know that Dennett has supposed there are gophers in Minnesota only by writing him, calling him long distance, waiting until we are in a symposium together and asking him, "Dan, do you suppose there are gophers in Minnesota?" and listening to what he says in response, and so on. Also I *could*, were it the Age of Utopian Neurophysiology, supplement the bases of my inferences to what he supposes with information concerning his neurophysiological states. (Of course it is only the *could* I care about, and I could not really care at all about what Dennett thought about the distribution of gophers.) The moral is that I know in my own case that and what I suppose without in any way being apprised of my own verbal output and/or neurophysiological states. But in the case of Dennett and/or any other "physical system," I could know about the existence and content of such a supposition only by an inference from verbal outputs and/or neurophysiological states. What, then, is the relationship between a mental act such as supposing (believing, thinking) and behavior and neurophysiological states? In short, the question of the relationship between the mental and the physical seems to remain extant.

Concerning closely related matters, Pylyshyn has written:

> Ultimately, the question of the nature of human intelligence may even involve us in a consideration of such *strong* con-

> straints as how a mechanism could realize such functions within the real-time and real-space confines of the brain, how it can do this subject to such principles as captured roughly in Lashely's notions of "equipotential" and "mass action," and finally how subjective experience itself is mapped onto these mechanisms.[66]

But something like this: the hope that perhaps with enough complexity and flexibility in a system consciousness might become ascribable to it, or that with a rich-enough set of extensional descriptions of a system, intentional descriptions would become appropriate to it, *or* that somehow, given who-knows-what neuro-cybernetic advance, subjective experience could be mapped onto mechanisms—all these are of the same cybernetic faith, and, at the moment, none of them take us beyond that.

But what would take us beyond such faith? The answer, I think, is the same for all the above positions: whatever it is which could exhibit a first-person perspective in essentially third-person sets of description. But the hope of being able to do that seems now to me a little less than the hope Abraham had for explaining to his family (or anyone else) why he was willing to sacrifice Isaac on Mt. Moriah.

John Haugeland remarked that "cognitive science sheds virtually no light on the issue of what consciousness is," but went on to say:

> So it's natural to suspect that something difficult and important is being left out. Unfortunately, nobody else has anything very specific or explanatory to say about consciousness either—it is just mysterious, regardless of your point of view. But that means that a cognitivist can say, "Look, none of us has much of an idea of what consciousness is; so how can we be so sure *either* that genuine understanding is impossible without it, *or* that semantic engines won't ever have it (e.g., when they are big and sophisticated enough)?" Those ques-

[66] Pylyshyn in Haugeland, *op. cit.*, p. 91.

tions may seem intuitively perverse, but they are very difficult to answer.[67]

But the problem of consciousness is, I think, even more perverse than Haugeland's remarks would indicate. For example, "big and sophisticated enough" at best goes proxy for something as vague as "adequate for consciousness." For when we consider it from a third-person point of view, there is a sense in which we seem always to come up empty handed. And this would still be so even if we were to imagine being graced with an exhaustive account of the processes underlying consciousness by Utopian neurophysiologists or robotologists. It is like the following example of Leibniz's:

> And supposing there were a machine, so constructed as to think, feel, and have perception, it might be conceived as increased in size, while keeping the same proportions, so that one might go into it as into a mill. That being so, we should, on examining its interior, find only parts which work one upon another, and never anything by which to explain a perception.[68]

Similarly, what would it be like either to see, or to comprehend without seeing, perceptual experience, or any other form of consciousness, within an AI model? On the other hand, when consciousness is considered from a first-person point of view, there also seems to be sense in which we feel almost omniscient with respect to what it is. We say, with great confidence, "Consciousness is the sort of experience I am now living through! What more could we want in order to understand it!?" But the answer that haunts us seems to be both "Nothing" and "Everything."

In connection with such an equivocal and wholly unsatisfactory answer, the robot whose misgivings about the

[67] Haugeland in Haugeland, *op. cit.*, p. 32.

[68] Leibniz, *The Monadology and Other Philosophical Writings*, trans. Robert Latta (London, 1951), pp. 227–28.

very possibility of modeling subjectivity prefaced this postscript had this to add:

> In spite of and rather independent of my otherwise Kierkegaardian propensities, I'd like to make a couple of more purely methodological and mundane observations concerning what I see as a kind of antinomy emerging from these matters. *On the one hand*, one feels that without addressing the mind-body relationship, neither strong AI *human* or weak AI *sim. human* could possibly succeed, and also that if such projects address the mind-body problem they couldn't possibly succeed. And *on the other hand*, one feels that to tender such criticism is to portray either or both as really full-fledged yet covert metaphysical enterprises as opposed to the empirical-theoretical investigations they are overtly and unabashedly proposed as being and in many respects actually seem to be. So although it can be admitted all around the town that the absence of a non-trivial representation of consciousness in a machine model of the mind constitutes a glaring gap, it also seems too short, sweet, and easy simply to point to that gap and suggest that because of it such models should be jettisoned. What perhaps needs to be done is to clarify the metaphysical limits of AI projects, and then address oneself to the topic of whether what lies outside those limits can be grappled with in some other way. It could, of course, turn out that what AI projects should be interpreted as being after all is just a lively sequence of *Gedankenexperiments*. Dennett (the post-*Brainstorms* Dennett) has suggested something along these lines in his "The Logical Geography of Computational Approaches (A View From the East Pole)" which I've only seen in a mimeographed version and can't quote without his permission.[69] And if he's right, then, of course, it may be that the limits of machine modeling of the mind coincide with the limits of human thought *simpliciter*, so that if a solution to the

[69] The robot seemed to me something of a literalist about this. I asked him to phone Dennett and request a juicy quote or two from this weird but interesting document, but he turned out to be too lazy ("tired" he said, contra Ziff, Chapter Three) to do so.

problem of non-trivially representing consciousness in a machine lies beyond that, it lies "way out there" in some ethereal realm beyond the reach of human (and I might add to include myself, robotic) comprehension.

Yet it may be that some sort of division of conceptual labor would be helpful at this point: i.e., a more careful delineation of the *kind* of contributions that AI projects might be expected to make to a characterization or taxonomy of the mind *short of dealing with the mind-body problem*. To do this, however, one needs an accurate etiology of the mind-body problem itself as well as its epistemic sidekick, the problem of other minds.[70] And, as is well known, philosophers are notoriously at odds with each other over what, in fact, the mind-body problem actually is! But whatever it is, it may be that the metaphysical therapy it cries out for could be solicited from explanatory sources other than those typically relied on in the invention and assessment of projects in AI. In other words, one might agree with Leibniz that we would never see or perceive a perception—*anywhere*! whether in some gargantuan machine or in an enlargement of micro-states and processes of the human brain—yet go on from there and explore why we wouldn't and what the implications of that are for how we might construe the mind-body relationship. And this latter exploration we might find can be indulged in without utilizing anything like the methodological tactics employed in AI, and without impugning those tactics either, except insofar as they purport to explain in programming terms aspects of our lives that are essentially non-programmable.

[70] For my own account of this, see "Asymmetries and Mind-Body Perplexities" (previously mentioned); "*Content and Consciousness* and the Mind Body Problem," *The Journal of Philosophy*, LXIX, No. 18 (Oct. 5, 1972), pp. 591–604; especially Section III, pp. 595–97; and "The Texture of Mentality" in *Wisdom—Twelve Esays*, ed. R. Bambrough (Oxford, 1974), pp. 173–93.

Selected Bibliography to the First Edition

Anderson, A., ed., *Minds and Machines,* Englewood Cliffs, N.J., 1964.

Armer, P., "Attitudes Toward Intelligent Machines," in E. Feigenbaum and J. Feldman, eds., *Computers and Thought,* New York, 1963.

Benacerraf, P., "God, the Devil and Gödel," *The Monist,* 51, 1967, 9–32.

Borko, H., ed., *Computer Applications in the Behavioral Sciences,* Englewood Cliffs, N.J., 1962.

Chomsky, N., *Syntactic Structures,* The Hague, 1956.

Cohen, L. J., "Can There Be Artificial Minds?" *Analysis,* 16, 1955, 36–41.

Descartes, R., *Philosophical Works,* trans. E. Haldane and G. R. T. Ross, New York, 1955.

Driesch, H., *The History and Theory of Vitalism,* New York, 1914.

Dreyfus, H., *Alchemy and Artificial Intelligence* (RAND Publication P-3244), Santa Monica, Calif., 1965.

Feigenbaum, E., and J. Feldman, eds., *Computers and Thought,* New York, 1963.

Feldman, J., "Computer Simulation of Cognitive Processes," in H. Borko, ed. *Computer Applications in the Behavioral Sciences,* Englewood Cliffs, N.J., 1962, pp. 337–56.

Fodor, J., and J. Katz, eds., *The Structure of Language: Readings in the Philosophy of Language,* Englewood Cliffs, N.J., 1964.

Gunderson, K., "Cybernetics," in P. Edwards, ed., *Encyclopedia of Philosophy,* New York, 1967.

—— "Cybernetics and Mind-Body Problems," *Inquiry,* 12, 1969, 406–19.

—— "Interview with a Robot," *Analysis,* 23, 1963, 136–42.

—— "Minds and Machines: A Survey," in R. Klibansky, ed., *Contemporary Philosophy: A Survey,* Florence, 1968.

—— "Asymmetries and Mind-Body Perplexities," in S. Winokur and M. Radner, eds., *Minnesota Studies in the Philosophy of Science,* Vol. IV, Minneapolis, 1970, pp. 273–309.

Huxley, T. H., "On the Hypothesis That Animals Are Automata, and Its History," in *Essays,* Vol. 1: *Methods and Results,* New York, 1911.

La Mettrie, J., *Man a Machine,* trans. by G. Bussey, Chicago, 1953.

Lange, F. A., *The History of Materialism,* trans. by E. C. Thomas, 3rd ed., London, 1950.

Lucas, J., "Minds, Machines, and Gödel," in A. Anderson, ed., *Minds and Machines,* Englewood Cliffs, N.J., 1964, pp. 43–59.

McDougall, W., *An Introduction to Social Psychology,* London, 1908; New York, 1961.

Miller, G., E. Galanter, and K. Pribram, *Plans and the Structure of Behavior,* New York, 1960.

Minsky, M., *"Steps towards Artificial Intelligence,"* in E. Feigenbaum and J. Feldman, eds., *Computers and Thought,* New York, 1963, pp. 406–56.

Newell, A., and H. Simon, "Computers in Psychology," in R. Luce, R. Bush, and E. Galanter, eds., *Handbook to Mathematical Psychology,* New York, 1963.

Parkinson, G. H., "The Cybernetic Approach to Aesthetics," *Philosophy,* 36, 1961, 49–61.

Puccetti, R., "On Thinking Machines and Feeling Machines," *British Journal for the Philosophy of Science,* 18, 1967, 39–51.

Putnam, H., "Minds and Machines," in A. Anderson, ed., *Minds and Machines,* Englewood Cliffs, N.J., 1964, pp. 72–97.

—— "The Mental Life of Some Machines," in H. N. Castaneda, ed., *Intentionality, Minds, and Perception,* Detroit, 1967, pp. 177–200.

—— "Robots: Machines or Artificially Created Life?," *Journal of Philosophy,* 61, 21, 1964, 668–91.

Reitman, W., *Cognition and Thought,* New York, 1966.

Rosenfield, L., *From Beast-Machine to Man-Machine: Animal Soul in French Letters from Descartes to La Mettrie,* New York, 1941.

Ryle, G., *The Concept of Mind,* London, 1949.

Sayre, K., *Recognition: A Study in the Philosophy of Artificial Intelligence,* Notre Dame, Ind., 1965.

Scriven, M., "The Mechanical Concept of Mind," in A. Anderson, ed., *Minds and Machines,* Englewood Cliffs, N.J., 1964, pp. 31–42.

—— "The Compleat Robot: A Prolegomena to Androidology," in S. Hook, ed., *Dimensions of Mind,* New York, 1960, pp. 118–42.

Shaw, R., T. Halwes, and J. Jenkins, *The Organism as a Mimicking Automaton* (Center for Research in Human Learning, University of Minnesota), in mimeo, 1966.

Turing, A. M., "Computing Machinery and Intelligence," in A. Anderson, ed., *Minds and Machines,* Englewood Cliffs, N.J., 1964, pp. 4–30.

Vartanian, A., *Diderot and Descartes: A Study of Scientific Naturalism in the Enlightenment,* Princeton, N.J., 1953.

—— *La Mettrie's L'Homme Machine: A Study in the Origins of an Idea,* Princeton, N.J., 1960.

White, B., "Studies of Perception, in H. Borko, ed., *Computer Applications in the Behavioral Sciences,* Englewood Cliffs, N.J., 1962, pp. 280–305.

Wittgenstein, L., *Philosophical Investigations,* trans. G. E. M. Anscombe, New York, 1953.

Ziff, P., "The Feelings of Robots," in A. Anderson, ed., *Minds and Machines,* Englewood Cliffs, N.J., 1964, pp. 98–103.

BIBLIOGRAPHIES

Minsky, M., "A Selected Descriptor-indexed Bibliography to the Literature on Artificial Intelligence," *IRE Transactions on Human Factors in Electronics,* Vol. HFE-2, 1961, 39–55. Also in Feigenbaum and Feldman, *op. cit.,* pp. 453–523.

Simmons, P. L., and R. F. Simmons, "The Simulation of Cognitive Processes: An Annotated Bibliography," *IRE Transactions on Electronic Computers,* Vol. EC-10, 1962, 462–83; Vol. EC-11, 1962, 535–52.

Bibliography to the Postscript

Colby, Kenneth Mark, "Modeling a Paranoid Mind," *The Behavioral and Brain Sciences*, 4, 1981, 515–60.

Churchland, Paul, *Matter and Consciousness*, Cambridge, Mass., 1984.

Condillac, *Treatise on the Sensations* (1754), trans. by Geraldine Carr, Los Angeles, 1930.

Dennett, Daniel, *Content and Consciousness*, London, 1969.

———*Brainstorms*, Montgomery, V., 1978.

———"The Logical Geography of Computational Approaches (A View from the East Pole)," mimeo, 1984.

Dickie, George, *Art and the Aesthetic—An Institutional Analysis*, Ithaca, N.Y., 1974.

Dreyfus, Hubert L., *What Computers Can't Do*, New York, 1972; second ed., 1979.

———"From Micro-Worlds to Knowledge Representation: AI at an Impasse," in Haugeland, excerpted from the Introduction to the above entry.

Geach, Peter, *Mental Acts*, London, 1952.

Gunderson, Keith, "*Content and Consciousness* and the Mind-Body problem," *The Journal of Philosophy*, LXIX, No. 18, Oct. 5, 1972, 591–604.

———"The Texture of Mentality," in R. Bambrough, ed., *Wisdom—Twelve Essays*, Oxford, 1974, pp. 173–93.

———"Purposes and Poetry," in D. F. Gustafson and B. L. Tapscott, eds., *Body, Mind, and Method*, The Hague, 1979, pp. 203–24.

———"Paranoia Concerning Program-Resistant Aspects of

the Mind, and Let's Drop Rocks on Turing's Toes Again," *The Behavioral and Brain Sciences*, 4, No. 4, 1981, 537–39.

Haugeland, John, ed. *Mind Design*, Montgomery, Vt., 1981. This includes his "Semantic Engines: An Introduction to Mind Design," pp. 1–34, and items by Newell and Simon, Pylyshyn, Dreyfus, Dennett, and Searle discussed in the Postscript, along with others.

Holly, William John, "Program-Resistant Aspects of Knowing-That and Knowing How," unpublished Ph.D. Dissertation, University of California, Irvine, 1975.

Leibniz, Gottfried Wilhelm, *The Monadology and Other Philosophical Writings*, trans. by Robert Latta, London, 1951.

Locke, John, *An Essay Concerning Human Understanding*, 1690; Dover Edition, New York, 1959.

Moor, James H., "Three Myths of Computer Science," *British Journal of the Philosophy of Science*, 29, 1978, 213–22.

Nagel, Thomas, "What Is It Like to Be a Bat?" in his *Mortal Questions*, Cambridge, England, 1979.

Newell, Alan, and Herbert A. Simon, "Computer Science as Empirical Inquiry," *Communications of the Association for Computing Machinery*, 19, 1976, 113–26. (Also in Haugeland.)

Premack, David, *Intelligence in Ape and Men*, Hillsdale, N.J., 1976.

———, and Guy Woodruff, "Does the Chimpanzee have a Theory of Mind?" *The Behavioral and Brain Sciences*, 1, No. 4, 1978., 515–26.

Pylyshyn, Zenon, "Complexity and the Study of Human Intelligence," in Martin Ringle, ed., *Philosophical Perspectives in Artificial Intelligence*, Atlantic Highlands, N.J., 1979. (Also in Haugeland.)

Reed, Michael, "Cognition and Explanation: Artificial Intelligence and the Philosophy of Mind," unpublished Ph.D. Dissertation, University of Minnesota, 1984.

Savage, C. Wade, ed. *Perception and Cognition: Issues in the Foundations of Psychology*, Vol. IX of *Minnesota Studies in the Philosophy of Science*, Minneapolis, 1978.

Schank, Roger C., "The Primitive Acts of Conceptual Dependency" in *Theoretical Issues in Natural Language Processing*, Cambridge, Mass., June 10–13, 1975.

Searle, John, "Minds, Brains, and Programs," first published in *The Behavioral and Brain Sciences*, 1, 1980, 417–24. (Also in Haugeland.)

Sloman, Aaron, and Monica Croucher, "Why Robots Will Have Emotions," in Ninth International Joint Conference on Artificial Intelligence, proceedings, 1981, 197–202.

Weizenbaum, Joseph, "ELIZA—A Computer Program for the Study of Natural Language Communication between Man and Machine," *Communications of the Association for Computing Machniery*, 9, 1965, 36–45.

——*Computer Power and Human Reason*, San Fancisco, 1976.

Wilkes, K. V., "Functionalism, Psychology, and the Philosophy of Mind," in *Philosophical Topics* (formerly *Southwestern Journal of Philosophy*), XII, 1981, 147–67.

Willick, Marshall S., "Artificial Intelligence: Some Legal Approaches and Implications," *AI Magazine*, summer 1983, 5–16.

Winograd, Terry, "Understanding Natural Language," *Cognitive Psychology*, 1, 1972, 1–191; also published by Academic Press, New York, 1972.

Wittgenstein, Ludwig, *Wittgenstein, Lectures and Conversations on Aesthetics, Psychology, and Religious Belief*, ed. Cyril Barrett, Los Angeles, 1972.

Proper Name Index

MENTALITY AND MACHINES
SECOND EDITION
KEITH GUNDERSON

Keith Gunderson received B.A. degrees from Macalester College and Oxford University and his Ph.D. in philosophy from Princeton University. He has taught philosophy at Princeton and UCLA and, since 1967, at the University of Minnesota, where he is now professor and a member of the Minnesota Center for Philosophy of Science. His articles on minds and machines, the mind-body problem, and aesthetics have appeared in philosophical journals and anthologies. Gunderson edited *Language, Mind, and Knowledge* (Volume VII in the series Minnesota Studies in the Philosophy of Science), and has published four books of poetry. Some of his poems have been set to music by the composers Sydney Hodkinson, Libby Larsen, and Eric Stokes, and parts of his poem sequence *Tripping Over the Cat* appeared in *Cat Catalog—The Ultimate Cat Book* (Workman 1976) and have been choreographed by Linda Shapiro of the New Dance Ensemble.